3rd International Conference on Information Processing and Control Engineering (ICIPCE 2019)

IOP Conference Series: Materials Science and Engineering
Volume 630

Moscow, Russia
4 – 7 August 2019

ISBN: 978-1-5108-9834-9
ISSN: 1757-8981

Printed from e-media with permission by:

Curran Associates, Inc.
57 Morehouse Lane
Red Hook, NY 12571

Some format issues inherent in the e-media version may also appear in this print version.

Institute of Physics
Dirac House, Temple Back
Bristol BS1 6BE UK

Phone: 44 1 17 929 7481
Fax: 44 1 17 920 0979

techtracking@iop.org

Additional copies of this publication are available from:

Curran Associates, Inc.
57 Morehouse Lane
Red Hook, NY 12571 USA
Phone: 845-758-0400
Fax: 845-758-2633
Email: curran@proceedings.com
Web: www.proceedings.com

Table of contents

Volume 630

3rd International Conference on Information Processing and Control Engineering

4–7 August 2019, Moscow, Russian Federation

Accepted papers received: 28 August 2019
Published online: 24 October 2019

Preface

Papers

Control Engineering

Electronic Design Technology

IOP Conf. Series: Materials Science and Engineering **630** (2019) 011001 doi:10.1088/1757-899X/630/1/011001

PREFACE

The 3rd International Conference on Information Processing and Control Engineering (ICIPCE 2019) held in Moscow – the Capital of Russian Federation during August 04-07, 2019.

The main topic of the conference is information processing and control, but certainly not limited to this. Information processing in the modern world is of great importance, which presents in so many areas of engineering and daily life. Thus, such a conference has scientific and applied value.

Nowadays, technologies related to information processing and management are developing rapidly. Russia has a good position in this area, so the choice of venue for such a conference is well justified.

At the conference, various reports were presented on the designated topics, in particular, there were reports devoted to the space industry, traditionally well represented in Russia, programming, process automation and production. There were representatives of universities, companies and academic institutions. The geography of participants included Russia, China, Poland, Kazakhstan, and South Korea and so on.

The conference consisted of two sections: Control system, Information system.

This conference is an excellent opportunity for scientists from different countries to meet with each other and discuss common tasks in a particular area of their research. Scientific schools, which include various scientists from different countries, will receive excellent experience from the possibility of such communication among scientists.

Conference Committee Chair
Vsevolod V. Koryanov
Bauman Moscow State Technical University, Russia

IOP Conf. Series: Materials Science and Engineering **630** (2019) 011001 doi:10.1088/1757-899X/630/1/011001

STATEMENT OF PEER REVIEW

All papers published in this volume of *IOP Conference Series: Materials Science and Engineering* have been peer reviewed through processes administered by the proceedings Editors. Reviews were conducted by expert referees to the professional and scientific standards expected of a proceedings journal published by IOP Publishing.

Conference Committee Chairs

Yury Razoumny, RUDN University, Russia
Vsevolod V. Koryanov, Bauman Moscow State Technical University, Russia

Program Committee Chairs

Marina Polyakova, Nosov Magnitogorsk State Technical University, Russia
Maxim Bakaev, Novosibirsk State Technical University, Russia
Andrey Malikov, North-Caucasus Federal University, Russia (Institute of ICT)

International Technical Committees

B.I. Shakhtarin, Bauman Moscow State Technical University, Russia
M.P. Sychev, Bauman Moscow State Technical University, Russia
A.B. Krasovskiy, Bauman Moscow State Technical University, Russia
V.T. Kalugin, Bauman Moscow State Technical University, Russia
Maria Lapina, North-Caucasus Federal University, Stavropol, Russia (Institute of ICT)
Vyacheslav Petrenko,North-Caucasus Federal University, Stavropol, Russia (Institute of ICT)
Yury Zagorulko, Institute of Informatics Systems of Russian Academy of Sciences, Russia
Haibo Zhao, Tongling University, China
Danhe Chen, Nanjing University of Science and Technology, China
Hsien-I Lin, National Taipei University of Technology, Taiwan
Lihong Zheng, Charles Sturt University, Australia
Zahid Akhtar, University of Memphis, USA
Adrian Kampa, Silesian University of Technology, Poland
Sebastian Heil, Chemnitz Technical University, Germany
Paulo Batista, CIDEHUS.UÉ - Interdisciplinary Center of History, Cultures and Societies of the University of Évora, Portugal
Mohamed bin Hussein, Universiti Technologi Malaysia, Malaysia
Dirman Hanafi, Universiti Tun Hussein Onn Malaysia, Malaysia
Godfried Toussaint, New York University, United Arab Emirates
Ibragim Suleimenov, Almaty University of Power Engineering and Telecommunications, Kazakhstan
Akhat Bakirov, Almaty University of Power Engineering and Telecommunications, Kazakhstan
Abdul Rehman Abbasi,Pakistan Institute of Engineering & Applied Sciences, Pakistan
Mahdi Baradarannia, University of Tabriz, Tabriz, Iran

IOP Conf. Series: Materials Science and Engineering **630** (2019) 011003 doi:10.1088/1757-899X/630/1/011003

Peer review statement

All papers published in this volume of *IOP Conference Series: Materials Science and Engineering* have been peer reviewed through processes administered by the proceedings Editors. Reviews were conducted by expert referees to the professional and scientific standards expected of a proceedings journal published by IOP Publishing.

IOP Conf. Series: Materials Science and Engineering **630** (2019) 012001 doi:10.1088/1757-899X/630/1/012001

Research on Coplanar Orbit Maneuvering in the Context of on-Orbit Service

Vsevolod V Koryanov[1], **Budang Zhang**[1]

[1] Dynamics and flight control of rockets and spacecraft, Bauman Moscow State Technical University, 105005, 2-nd Baumanskaya, 5, b1, Moscow, Russian Federation

vkoryanov@mail.ru , zdang_ivan@163.com

Abstract. The concept of on-orbit service is introduced, and the development process of on-orbit service is briefly reviewed. The optimal transfer plan of orbit transfer and rendezvous is studied in the context of on-orbit service. The circular orbit motion with perturbation is used to study the near-circle orbit maneuver, and the general equation of coplanar maneuver is established based on the cylindrical coordinate system. The influences of the orbital parameters and the difference initial phase between the coaxial orbits on the optimal transfer orbit are calculated by Matlab simulation.

1. Introduction

The conceptual study of spacecraft on-orbit service began with the Soviet Union replenishment of the space station, followed by special attention in aerospace engineering such as the Soviet spacecraft and the US space shuttle [1]. NASA gave such a description of on-orbit service in the "NASA satellite servicing project report" – In space, people, robots or both work together to complete a type of operation involving the life of various spacecraft and improving the ability to perform tasks[2]. For nearly half a century, the United States, the Soviet Union, Japan, Canada, and ESA have conducted a series of research and experiments in conjunction with specific space missions[3]. In the past 20 years, around the application research of on-orbit service technology, the world's major space powers have carried out a series of exploratory attempts. For example, the US astronauts operated outside the space shuttle and successfully recovered several satellites such as the Palm House B2 and West Star 6, repaired the Hubble telescope, installed giant solar panels for the International Space Station, and installed the space station JEMRMS robotic arm, etc[3-5]. Simultaneously the number of space debris has significantly increased in the near-Earth space as a consequence of continued space activities.[6] Because the positive effect of atmosphere on space debris is very limited[7], active debris removal has become a hot research in on-orbit services.

In this context, the concept of a space-based orbital service system was proposed.[8] The operating system consists of three parts: "Autonomous space transport and orbital robot" (service station), "service vehicle" (active device) and "serviced vehicle" (space vehicle - target). On-orbit service trajectory information as shown in Fig. 1.

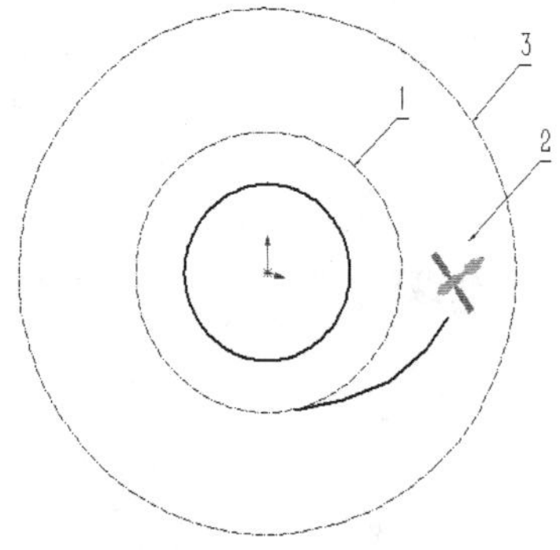

Figure 1. Principled scheme of the on-orbit service
1- Parking orbit "Service Station"
2- Active spacecraft
3-Target orbit "Passive spacecraft"

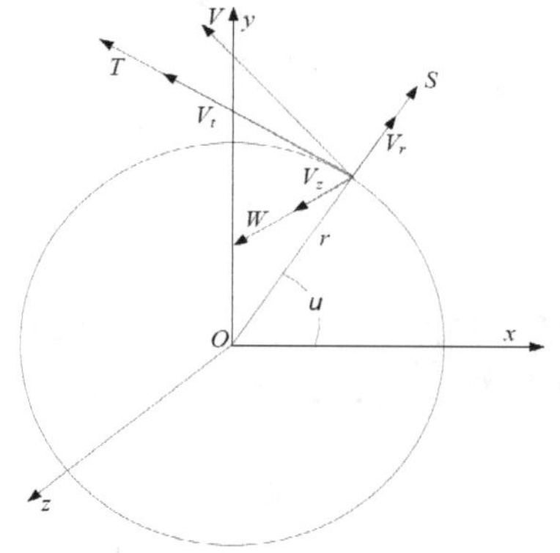

Figure 2. The cylindrical coordinate system

In this space-based on-orbit service system, the transfer and rendezvous of spacecraft is a critical step in completing on-orbit service. At present, only the United States, Russia and China independently master space rendezvous and docking technology. Based on the limitations of spacecraft fuels, active spacecraft should minimize fuel consumption during orbital transfer and rendezvous. Therefore, for different initial orbits and service orbits, the design of the optimal transfer orbit has become a research topic for many scholars.[9]

2. Equation of orbital maneuver

When studying the spacecraft maneuver near orbit, the near-circular orbit motion is regarded as the disturbed motion of the circular orbit, and the orbital maneuver equation is established on the cylindrical coordinate system r, u, z (Figure 2.). In this coordinate system, the established coplanar orbit maneuver equation is as in equation (1).[9-15]

$$\begin{gathered}\sum_{i=1}^{N}(\Delta V_{ri}\sin\varphi_i + 2\Delta V_{ti}\cos\varphi_i) = \Delta e_x \\ \sum_{i=1}^{N}(-\Delta V_{ri}\cos\varphi_i + 2\Delta V_{ti}\sin\varphi_i) = \Delta e_y \\ \sum_{i=1}^{N}2\Delta V_{ti} = \Delta a \\ \sum_{i=1}^{N}(2\Delta V_{ri}(1-\cos\varphi_i) + \Delta V_{ti}(-3\varphi_i + 4\sin\varphi_i)) = \Delta t\end{gathered}, \quad (1)$$

where, $\Delta e_x = e_f\cos\omega_f - e_0\cos\omega_0$, $\Delta e_y = e_f\sin\omega_f - e_0\sin\omega_0$, $\Delta a = (a_f - a_0)/r_0$, $\Delta t = \lambda_0(t_f - t_0)$, $\Delta V_{ri} = \Delta V_{ri}^* / V_0$, $\Delta V_{ti} = \Delta V_{ti}^* / V_0$.

It is necessary to clarify that this equation is a non-dimensional equation. Here «f», «0» – indices corresponding to the final and initial orbits, e_f, e_0 – eccentricity of the orbit; a_f, a_0 – semi-major axes of the orbits; ω_f, ω_0 – angles between the direction to the pericenter of the corresponding orbit and the direction to the point specified in the final orbit; t_f – the required time of arrival at a given point, t_0 – initial maneuver time; N – number of speed pulses; φ_i – the angle of application of the i-th velocity pulse, measured from the direction to a given point in the direction of the spacecraft; $\Delta V_{ri}^*, \Delta V_{ti}^*$ – radial and transversal components of the i-th velocity pulse, respectively.

The task of determining the optimal maneuver parameters for coplanar orbits is specified in the following manner: determine $\Delta V_{ri}, \Delta V_{ti}, \varphi_i (i = 1,..., N)$ to minimize the total feature speed of the orbital maneuver in (2).

$$\Delta V = \sum_{i=1}^{N} \left(\Delta V_{ri}^2 + \Delta V_{ti}^2 \right)^{1/2} \tag{2}$$

3. Impact of parameters on the optimal orbital maneuver

3.1. Impact of perigee and apogee on optimal orbital maneuver

According to formula (1), the velocity impulses of the spacecraft flight from apogee to apogee (T1) and from perigee to perigee (T2) are calculated separately in the coaxial orbit. And compare their size to determine the optimal orbit maneuver and study the influence of orbital parameters on it. Coaxial orbits have intersecting orbits and non-intersecting orbits, as shown in Figure 3.

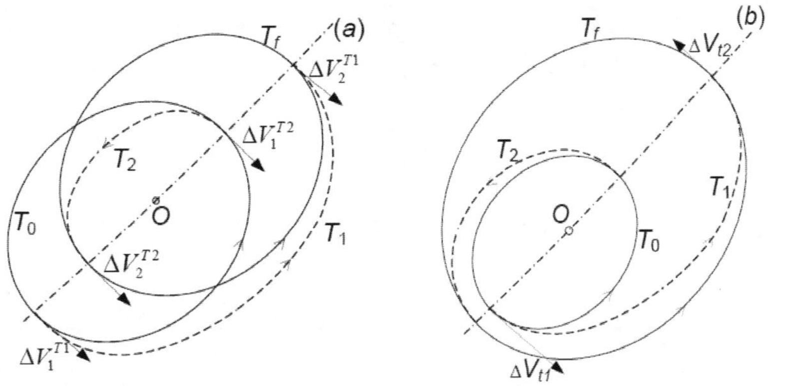

Figure 3. Transitions between reverse coaxial orbits:
(a)- between intersecting orbits; (b) - between non-intersecting orbits.

3.1.1. According to intersecting coaxial orbits. Change the altitude of the apogee of the orbit, calculate the velocity impulses required for the orbital transfer along the track T1 and the track T2, and obtain the table 1.

Table 1. The influence of apogee height on speed impulse

Orbital parameters	$h_p = 200km$ $h_a = 400km$	$h_p = 200km$ $h_a = 450km$	$h_p = 200km$ $h_a = 500km$	$h_p = 200km$ $h_a = 550km$
Eccentricity	0.0150	0.0187	0.0223	0.0259
$\Delta V_1^{T1} / (m/s)$	58.8118	73.3728	87.8769	102.3240
$\Delta V_1^{T2} / (m/s)$	-57.0746	-70.6836	-84.0401	-97.1494
Difference,%	2.9538	3.6652	4.3662	5.0571
$\Delta V_2^{T1} / (m/s)$	-56.6580	-70.0452	-83.1384	-95.9456
$\Delta V_2^{T2} / (m/s)$	59.2643	74.0803	88.8964	103.7125
Difference,%	4.6001	5.7608	6.9295	8.0951
$\Delta V_1 / (m/s)$	115.4698	143.4180	171.0153	198.2696
$\Delta V_2 / (m/s)$	116.3389	144.7640	172.9365	200.8619
Difference,%	0.7527	0.9385	1.1234	1.3075

When the perigee height is constant, the higher the altitude of the apogee, the greater the difference in velocity impulses required for the spacecraft to travel along the track T1 and the track T2. As shown in Table 1, the difference in individual speed pulses reaches several percentage points, while the total speed impulse is not much different, about one percentage point. Therefore, from a mathematical point of view, the optimal transfer orbit for fuel saving should be T1. From the engineering point of view, the T1 and T2 tracks are the best transfer orbit as long as the error is within a reasonable range.

Following the above process, when the apogee is unchanged, change the height of the perigee, and use the same method to get the table 2.

Table 2. The influence of perigee height on speed impulse

Orbital parameters	$h_p = 200km$ $h_a = 400km$	$h_p = 250km$ $h_a = 400km$	$h_p = 300km$ $h_a = 400km$	$h_p = 350km$ $h_a = 400km$
Eccentricity	0.0150	0.0112	0.0074	0.0037
$\Delta V_1^{T1} / (m/s)$	58.8118	43.6961	28.8596	14.2962
$\Delta V_1^{T2} / (m/s)$	-57.0746	-42.7281	-28.4334	-14.1906
Difference,%	2.9538	2.2153	1.4769	0.7384
$\Delta V_2^{T1} / (m/s)$	-56.6580	-42.4935	-28.3290	-14.1645
$\Delta V_2^{T2} / (m/s)$	59.2643	43.9457	28.9684	14.3228
Difference,%	4.6001	3.4175	2.2569	1.1180
$\Delta V_1 / (m/s)$	115.4698	86.1896	57.1886	28.4607
$\Delta V_2 / (m/s)$	116.3389	86.6738	57.4017	28.5135
Difference,%	0.7527	0.5618	0.3727	0.1855

As seen from Table 2, the higher the height of the perigee, the smaller the difference between the individual velocities pulses and the total velocity pulse. At this time, the difference between the individual velocities pulses and the total velocities pulses on the transfer track along T1 and T2 is small, so the track transfer can be achieved on the T1 and T2 tracks.

Based on the comprehensive analysis of Table 1 and Table 2, as the orbital eccentricity increases, the difference between the orbital transfer velocity pulses from T1 and T2 tracks becomes larger and larger. Therefore, when the eccentricity of the track is large enough, the transfer track T1 is selected as the optimum transfer trajectory. When the orbital eccentricity is small, the transfer window should be considered comprehensively.

3.1.2. According to non-intersecting coaxial orbits. Change the orbital parameters of the initial orbit and service orbit respectively, calculate according to the method of 3.1.1 to obtain Table 3. It can be clearly seen from Table 3 that the optimal transfer orbit should be the T2 track.

Table 3. Influence of orbital parameters on optimal maneuver

Initial orbit parameters	$h_{p1} = 200km$ $h_{a1} = 400km$	$h_{p1} = 200km$ $h_{a1} = 400km$	$h_{p1} = 200km$ $h_{a1} = 400km$	$h_{p1} = 400km$ $h_{a1} = 600km$	$h_{p1} = 600km$ $h_{a1} = 800km$
Service Orbit Parameters	$h_{p2} = 500km$ $h_{a2} = 600km$	$h_{p2} = 600km$ $h_{a2} = 800km$	$h_{p2} = 800km$ $h_{a2} = 1000km$	$h_{p2} = 800km$ $h_{a2} = 1000km$	$h_{p2} = 800km$ $h_{a2} = 1000km$
$\Delta V_1^{T1}, (m/s)$	411.6826	529.3062	646.9298	618.6202	592.3166
$\Delta V_1^{T2}, (m/s)$	228.2985	285.3731	399.5224	382.3729	366.4162
$\Delta V_2^{T1}, (m/s)$	-227.5388	-246.0116	-239.4799	-318.3964	-389.3904

$\Delta V_2^{T2},(m/s)$	-81.3438	-53.4547	-51.9561	-147.6070	-233.3908
$\Delta V_1,(m/s)$	639.2213	775.3177	886.4097	937.0166	981.7070
$\Delta V_2,(m/s)$	309.6423	338.8279	451.4784	529.9799	599.8070

3.2. The effect of initial phase on optimal orbital maneuver

For spacecraft intersection problems, when the intersection time is given, different initial phases will affect the optimal transfer orbit. Table 4 gives the initial orbit and service orbit information and gives the initial phase and orbit intersection information.

Table 4. Orbital parameters and rendezvous information

Parameters	Initial orbit	Service orbit
$H_{\min}(km)$	180	401
$H_{\max}(km)$	210	408
$\omega_{prg}(degree)$	20	102.7
$U_1(degree)$	60	5
$U_2(degree)$	60	210
$U_3(degree)$	60	355
$N_0-Number\ of\ initial\ turns$	1	201
$N_{ap}-Timing\ time$	17	217

The simplest apsidal solution will be used[9], and according to the solution result draw the transfer orbit for different initial phases, as shown in Figure 4.

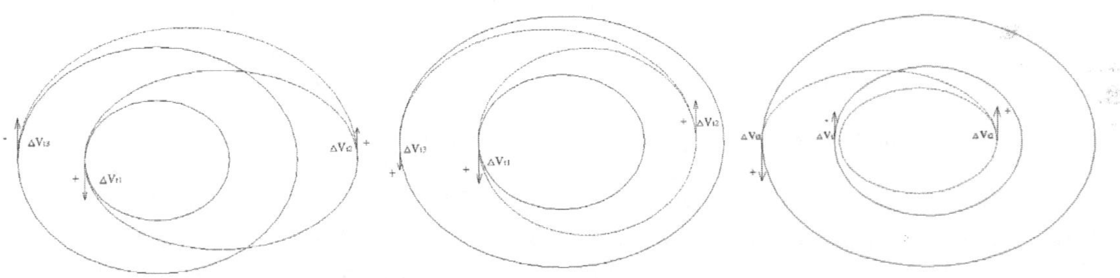

a) $U_{\kappa 1}=5^{\circ}$ b) $U_{\kappa 2}=210^{\circ}$ c) $U_{\kappa 3}=355^{\circ}$

Figure 4. Track transfer scheme with different initial phases

In the first case (Figure 4, a), the target spacecraft lags behind the active spacecraft. In order to reduce this lag, it is necessary to significantly increase the semi-major axis of the phase-modulated orbit, which will allow the target spacecraft moving in orbit with a smaller period to "catch up" the active spacecraft. This type of flight is not optimal because of the brake maneuver. On the other hand, due to the number of laps in flight, it is impossible to apply a reverse scheme to make the active spacecraft "catch up" with the target spacecraft.

In the second case (Figure 4, b), the active spacecraft is located behind the target spacecraft. The phase difference is within a reasonable range, and the three velocity pulses obtained are all acceleration pulses. This maneuvering scheme is the optimal maneuver scheme.

In the third case (Figure 4, c), the active spacecraft lags far behind the target spacecraft. In order to catch up with the target spacecraft within a given intersection time, it is necessary to further reduce the orbital height to obtain a track period with a small angle. Therefore, this transfer scheme has a brake pulse, which is not the optimal transfer scheme.

Based on the above analysis, it is found that the difference of the initial phase will cause the occurrence of the brake pulse during the determined intersection time. At this time, the maneuver scheme solved by the simplest apsidal solution is not the optimal maneuver scheme. Observing the calculation process, it is found that the scalar sum is added between the various speed increments in the simplest apsidal solution , which results in the limitation of satisfying the minimum total speed impulse when a certain speed is the braking speed.

4. Conclusions

For the coaxial orbit, the general equations established by the application find that when the initial orbit intersects with the service orbit, the velocities pulses of the active spacecraft from the apogee to the apogee and from the perigee to the perigee is not much different. However, when the orbital eccentricity is large enough, it is recommended to select a transfer orbit from apogee to apogee within a reasonable intersection window. When the initial orbit is not intersected with the service track, the transfer orbit from the perigee to the perigee is the best orbit. When the target spacecraft lags behind the active spacecraft given the constraints of the Rendezvous time, or the target spacecraft is far ahead of the active spacecraft, the transfer orbit solved by the simplest apsidal solution is not the optimal transfer orbit. For the above two cases, the solution of the optimal orbit needs further study.

Acknowledgements

This work was supported by the China Scholarship Council (No.201806030227). The authors thank the China Scholarship Council for awarding a fellowship.

References

[1] Hu S, Li Y and Chen X 2016 J. *Manned Spaceflight* 22 452-458
[2] NASA 2010 *On-Orbit Satellite Servicing Study* p 179
[3] Chen X, Yuan J and Yao W Technologies of Spacecraft On-orbit Service (Beijing: China Aerospace Press) p 120
[4] Hecht J R 1995 J. Laser Focus World 31 57
[5] Wen Q, Ma X and Hua Y 2013 J. *Manned Spaceflight* 19 86-90
[6] Qi R, Misra A K and Zuo Z 2017 J. *Journal of Guidance, Control and Dynamics* 40 720-728
[7] Bonnal C, Ruault J M and Desjean M C 2013 J. *Acta Astronautica* 85 557-569
[8] Li Y and Cheng L 2008 J. *Missile and Vehcile* 5 31-34
[9] Baranov A A Maneuvering of spacecraft in the vicinity of a circular orbit (Moscow: Satellite +) p 512
[10] Baranov A A, Grishko D A and Razoumny Y N 2017 J. *Cosmic Research* 55 361-370
[11] Baranov A A and Grishko D A 2017 J. *Advances in the Astronautical Sciences* 161 519-526
[12] Baranov A.A., Grishko D.A., Medvedevskikh, V.V., Lapshin, V.V. Solution of the flyby problem for large space debris at sun-synchronous orbits. Cosmic Research, Volume 54, Issue 3, 1 May 2016, Pages 229-236.
[13] Baranov A.A., Grishko D.A. Optimization of flyby schemes in the framework of ADR mission in LEO. Advances in the Astronautical Sciences Volume 161, 2017, Pages 519-526. 3rd International Academy of Astronautics Conference on Dynamics and Control of Space Systems, DyCoSS 2017; Moscow; Russian Federation; 30 May 2017 - 1 June 2017; Code 213319.
[14] Vsevolod Koryanov, Da-Poian Victoria. Space servicing: Strategy and logistics // Proceedings of the International Astronautical Congress, IAC, 2018-October
[15] Mayorova V.I., Leonov, V., Grishko, D.A. Probable approaches to the near-earth orbits clean-up from space debris with dimensions less than 10 centimeters. Proceedings of the International Astronautical Congress, IAC. Volume 4, 2015, Pages 2719-2722, 66th International Astronautical Congress 2015: Space - The Gateway for Mankind's Future, IAC 2015; Jerusalem; Israel; 12 October 2015 to 16 October 2015; Code 122921.

IOP Conf. Series: Materials Science and Engineering **630** (2019) 012002 doi:10.1088/1757-899X/630/1/012002

Postural Stabilization of Quadrotor using Extended Kalman Filter and Integral Sliding Mode Control

Hyukwoo Lee[1], Kyunghyun Lee[1], and Kwanho You[1,*]

[1]Department of Electrical and Computer Engineering, Sungkyunkwan University, Suwon, Korea

[*]corresponding author: khyou@skku.edu

Abstract. In this paper, we consider the stabilization and the estimate of quadrotor's posture. Nowadays, as the interest in unmanned aerial vehicle (UAV) has increased, various application fields using UAV have appeared. The quadrotor is the most common type of UAV and many approaches to control the system have been studied. In real control environment, the system of quadrotor is affected by disturbance and measurement noise. Minimizing the effects of disturbance or measurement noise is important part. We propose the extended kalman filter (EKF) to reduce effectively the measurement noise of the nonlinear drones system and the integral sliding mode control (ISMC), a robust controller for the model uncertainty and disturbance. We show the proposed control performance through some simulations.

1. Introduction

The quadrotor is the most general type of UAV, lifted by four motors and propellers. The system of quadrotor has advantages from a variety of applications due to simple structure, vertical take-off and landing (VTOL) capability and payload [1]. Each motor is oriented vertically downward, with two motors rotating in clockwise and the other two motors rotating in counterclockwise. By controlling the four motor speeds, the attitude and altitude of quadrotor can be adjusted. The quadrotor has rolling, pitching and yawing corresponding to the x, y, and z axis, respectively.

Several control techniques such as proportional-integral-differential (PID), backstepping, and optimal control have been studied for controlling the quadrotor system. Garcia [2] proposed a robust PID control strategy via affine parametrization designed for multivariable nonlinear quadrotor. The proposed PID control system is assured by using the H_∞ norm of the weighted complementary sensitivity function. Satici [3] provided the design and implementation of an L_1 -optimal control of the quadrotor dynamical system that rejects disturbances. The proposed algorithm reduces the magnitude of the errors in L_1 -optimal sense.

In this paper, we propose ISMC and EKF to reduce the effects of measurement noise and to design a robust controller for disturbance and model uncertainty. This paper organized as follows. In section II, we obtain the dynamic model of quadrotor and show the overall flow of quadrotor system with EKF and ISMC. In section III, we show the composition and derive the process of EKF. In section IV, we obtain the control input by applying the EKF angle estimates and the dynamic model of the quadrotor to the ISMC. In section V, to prove the performance the stabilization and the estimate, we show simulation results by comparing with a general SMC. Finally, section VI presents the conclusion.

2. Model of quadrotor's dynamic

Quadrotor is a nonlinear system with many state variables [4]. The main forces acting on the quadrotor are generated by four motors and the propellers. The quadrotor consists of a fixed body frame and the inertial frame *E*. The four motors are coupled with each propeller, which intersects with each other. To balance the quadrotor, one pair of motors (w_2, w_4) rotate clockwise, and the other motors (w_1, w_3) rotate counterclockwise. We can change the direction and the position of a quadrotor by varying the speed of the four propellers.
The dynamic model of the quadrotor can be derived by Euler-Largrange [5] as follows:

$$\begin{aligned} \ddot{\phi} &= \left(\frac{i_y - i_z}{i_x}\right)\dot{\theta}\dot{\psi} + \left(\frac{L}{i_x}\right)u_1, \\ \ddot{\theta} &= \left(\frac{i_z - i_x}{i_y}\right)\dot{\phi}\dot{\psi} + \left(\frac{L}{i_y}\right)u_2, \\ \ddot{\psi} &= \left(\frac{i_x - i_y}{i_z}\right)\dot{\phi}\dot{\theta} + \left(\frac{L}{i_z}\right)u_3. \end{aligned} \quad (1)$$

ϕ, θ and ψ are the angles of the quadrotor rotation about the *x, y, z* axis and are the roll, pitch, yaw angles, respectively. i_x, i_y and i_z are the body inertia of a quadrotor and L is the lever length. u_i $(i = 1,2,3)$ is the control input of a quadrotor system.

$$\begin{aligned} u_1 &= b\left(\omega_4^2 - \omega_2^2\right), \\ u_2 &= b\left(\omega_3^2 - \omega_1^2\right), \\ u_3 &= d\left(\omega_2^2 + \omega_4^2 - \omega_1^2 - \omega_3^2\right), \\ \omega &= \omega_2 + \omega_4 - \omega_1 - \omega_3. \end{aligned} \quad (2)$$

Where b and d represent the thrust factor and the drag factor, respectively. ω_i $(i = 1,2,3,4)$ is the rotational speed of i-th motor. From equation (1), the model of a quadrotor can be rewritten in a state space form as follows:

$$\begin{aligned} &x_1 = \phi,\ x_3 = \theta,\ x_5 = \psi, \\ &x_2 = \dot{x}_1 = \dot{\phi},\ x_4 = \dot{x}_3 = \dot{\theta},\ x_6 = \dot{x}_5 = \dot{\psi}, \end{aligned} \quad (3)$$

From equations (1) and (3), the quadrotor dynamic is formulated as:

$$\dot{\mathbf{x}} = \mathbf{m}(\mathbf{x},\mathbf{u}) = \begin{pmatrix} x_2 \\ a_1 x_4 x_6 + b_1 u_1 \\ x_4 \\ a_2 x_2 x_6 + b_2 u_2 \\ x_6 \\ a_3 x_2 x_4 + b_3 u_3 \end{pmatrix} \quad (4)$$

where

$$\begin{aligned} &a_1 = \frac{i_y - i_z}{i_x},\ a_2 = \frac{i_z - i_x}{i_y},\ a_3 = \frac{i_x - i_y}{i_z}, \\ &b_1 = \frac{L}{i_x},\ b_2 = \frac{L}{i_y},\ b_3 = \frac{L}{i_z}, \end{aligned} \quad (5)$$

3. State estimation using EKF

Kalman filter is a method of estimating the state of linear system that contains measurement noise or error [6]. Kalman filter basically assumes the linearity of the model, however in practice many models are nonlinear structures. Thus, the continuous-discrete EKF can be applied to estimate states.

The algorithm consists of two steps: time update and measurement update. The nonlinear discrete system with measurement noise is given as follows:

$$\begin{aligned} \dot{\mathbf{x}} &= \mathbf{m}(\mathbf{x},\mathbf{u},t)+\boldsymbol{\Theta}(t)w, \\ \mathbf{z_k} &= \mathbf{y}\left[\mathbf{x}(t_k),k\right]+v_k, \end{aligned} \tag{6}$$

where $\mathbf{m}(\mathbf{x},\mathbf{u},t)$ is a model of quadrotor system and $\mathbf{y}\left[\mathbf{x}(t_k),k\right]$ is a measurement model, respectively. The quadrotor is a time varying system and the sensor in a quadrotor measures in discrete time. In equation (6), $\boldsymbol{\Theta}(t)$ is the process noise matrix. $w(t) \approx (0,\ Q)$ and $v_k \approx (0,\ R)$ are the white noises uncorrelated with each other and independent from the system state.

$$\mathbf{M}(\mathbf{x},t)=\frac{\partial \mathbf{m}(\mathbf{x},\mathbf{u},t)}{\partial \mathbf{x}}\ ,\ \ \mathbf{Y}(\mathbf{x})=\frac{\partial \mathbf{y}(\mathbf{x},k)}{\partial \mathbf{x}}, \tag{7}$$

where $\mathbf{M}(\mathbf{x},t)$ and $\mathbf{Y}(\mathbf{x})$ are Jacobians derived from $\mathbf{m}(\mathbf{x},\mathbf{u},t)$ and $\mathbf{y}\left[\mathbf{x}(t_k),k\right]$, respectively. The first step of the Kalman filter is time update. To predict the next state, we use the priori measurement update state.

$$\begin{aligned} \dot{\hat{\mathbf{x}}} &= \mathbf{m}(\hat{\mathbf{x}},\ \mathbf{u},\ t), \\ \dot{\mathbf{P}} &= \mathbf{M}(\hat{\mathbf{x}},\mathbf{t})\mathbf{P}+\mathbf{P}\mathbf{M}^{\mathrm{T}}(\hat{\mathbf{x}},\mathbf{t})+\boldsymbol{\Theta}Q\boldsymbol{\Theta}^{T}. \end{aligned} \tag{8}$$

$\mathbf{P}$ is the error covariance matrix obtained from the prior measurement update. To estimate measurement, the gain $\mathbf{K_k}$ of Kalman filter is obtained as follows:

$$\mathbf{K_k}=\mathbf{P^-}(\mathbf{t_k})\mathbf{Y^T}(\hat{\mathbf{x}}_\mathbf{k}^-)\left[\mathbf{Y}(\hat{\mathbf{x}}_\mathbf{k}^-)\mathbf{P^-}(\mathbf{t_k})\mathbf{Y^T}(\hat{\mathbf{x}}_\mathbf{k}^-)+\mathbf{R}\right]^{-1}. \tag{9}$$

Using the Kalman filter gain obtained in equation (9) and the error covariance matrix of the previous state, $\mathbf{P}(\mathbf{t_k})$ and measurement estimate are updated as follows:

$$\begin{aligned} \mathbf{P}(\mathbf{t_k}) &= \left[\mathbf{I}-\mathbf{K_k}\mathbf{Y}(\hat{\mathbf{x}}_\mathbf{k}^-)\mathbf{P^-}(\mathbf{t_k})\right], \\ \hat{\mathbf{x}}_\mathbf{k} &= \hat{\mathbf{x}}_\mathbf{k}^-+\mathbf{K_k}\left[\mathbf{z_k}-\mathbf{Y}(\hat{\mathbf{x}}_\mathbf{k}^-,\mathbf{k})\right], \end{aligned} \tag{10}$$

where $\mathbf{I}$ is an identity matrix. From equation (10), the update of $\mathbf{P}(\mathbf{t_k})$ and $\hat{\mathbf{x}}_\mathbf{k}$ is derived and these values are applied to the covariance error matrix of time update and measurement model. The EKF can be applied to the quadrotor dynamic system, we can estimate the angles of the quadrotor..

4. ISMC for quadrotor stabilization

ISMC is not sensitive to modeling errors and parameter uncertainties with disturbances [7]. To obtain the control input of a quadrotor, we use ISMC with the estimated angles obtained from EKF. The sliding surface is designed as follows:

$$s_1=(\dot{\phi}_d-\dot{\phi})+\lambda_1(\phi_d-\phi)+\delta_1\int_0^t(\phi_d-\phi)d\tau. \tag{11}$$

In equation (11), λ_1 and δ_1 are positive constants of a sliding surface. To satisfy the stability condition of ISMC, we use the Lyapunov function [8].

$$V(s_1) = \frac{1}{2}s_1^{\ 2},\ \dot{V}(s_1) = s_1\dot{s}_1 < 0. \tag{12}$$

To satisfy the Lyapunov conditions (12) and (13), we differentiate the sliding surface using equations (4) and (11) as follows:

$$\begin{aligned} \dot{s}_1 &= -k_1\, sat(s_1) - k_2 s_1 \\ &= \ddot{\phi}_d - \left(a_1\dot{\theta}\dot{\psi} + b_1 u_1\right) + \lambda_1\left(\dot{\phi}_d - \dot{\phi}\right) + \delta_1(\phi_d - \phi). \end{aligned} \tag{13}$$

To avoid the chattering which happens in ISMC, we replaced the signum function with a saturation function. With equations (14) and (15), we can derive the control input u_i as

$$u_1 = \frac{1}{b_1}\left[\ddot{\phi}_d - a_1\dot{\theta}\dot{\psi} + \lambda_1\left(\dot{\phi}_d - \dot{\phi}\right) + \delta_1(\phi_d - \phi) + k_1 sat\left(s_1\right) + k_2 s_1\right]. \tag{14}$$

Similarly, the control inputs can be derived as:

$$\begin{aligned} u_2 &= \frac{1}{b_2}\left[\ddot{\theta}_d - a_2\dot{\phi}\dot{\psi} + \lambda_2\left(\dot{\theta}_d - \dot{\theta}\right) + \delta_2\left(\theta_d - \theta\right) + k_3 sat\left(s_2\right) + k_4 s_2\right], \\ u_3 &= \frac{1}{b_3}\left[\ddot{\psi}_d - d_5\dot{\phi}\dot{\theta} + \lambda_3\left(\dot{\psi}_d - \dot{\psi}\right) + \delta_3\left(\psi_d - \psi\right) + k_5 sat\left(s_3\right) + k_6 s_3\right]. \end{aligned} \tag{15}$$

5. Simulation and analysis

In this section, in order to evaluate the estimate performance of the continuous-discrete EKF and stabilization of ISMC, we analyze the simulation result. In simulations, the measurement state of quadrotor consists of estimated values by EKF. Figure 1 shows the estimate values and the stabilizations of a quadrotor about roll.

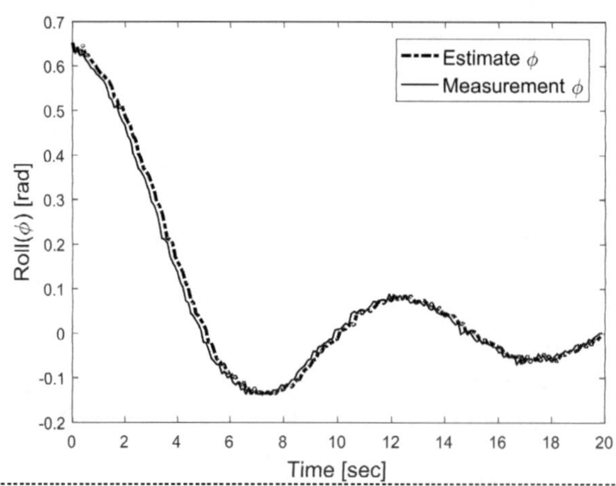

Figure 1. Estimate and measurement of roll angle.

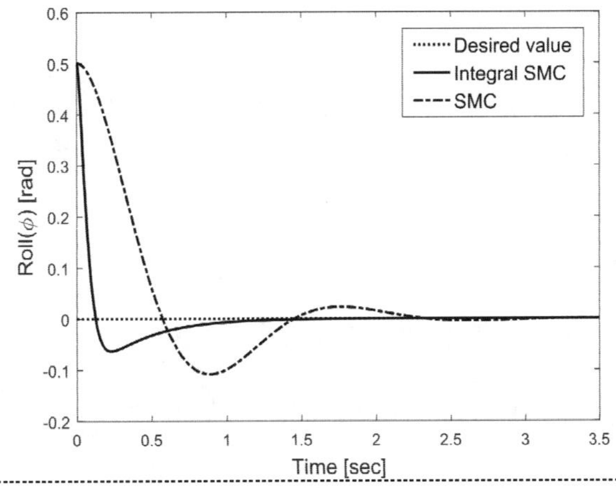

Figure 2. Stabilization comparison between ISMC and SMC.

Figure 2 shows the control performance comparison between the ISMC and the general SMC. Simulation was carried out to stabilize the roll angle, and the initial values of ISMC and SMC are set as 0.5 radian. The settling times of ISMC and SMC are 0.86 sec and 1.4 sec, respectively.

6. Conclusion

In this paper, we proposed a postural stabilization of a quadrotor using ISMC and EKF. We applied a continuous-discrete EKF to estimate the angle from the discontinuous features of sensor and the continuous system of the quadrotor. We applied the estimate values obtained from EKF to the ISMC for reducing the measurement error of a sensor. The estimation and the control performance of our proposed method have been proved through the simulations.

Acknowledgments

This work was supported by the National Research Foundation of Korea (NRF) grant funded by the Korea government (MSIP) (NRF-2019R1A2C1002343).

References

[1] Tayebi A and McGilvray S 2006 Attitude stabilization of a VTOL quadrotor aircraft *IEEE Transactions on Control Systems Technology*. vol. 14, pp. 562-571

[2] Garcia R and Rubop F 2012 Robust PID control of the quadrotor helicopter *Proceeding of 18th World Congress IFAC*. vol. 45, pp. 229-234

[3] Satici A, Poonawala H, and Spong M 2013 Robust optimal control of quadrotor UAVs *IEEE Access*. vol. 1 pp. 79-93

[4] Selfridge J and Tao G 2013 A multivariable adaptive controller for a quadrotor with guaranteed matching conditions *Systems Science & Control Engineering*. vol. 2, pp. 24-33

[5] Naidoo Y and Stopforth R 2011 Quad-rotor unmanned aerial vehicle helicopter modeling & control *International Jounal of Advanced Robotic Systems*. vol. 8, pp. 139-149

[6] Evensen 2003 The ensemble Kalman filter: theoretical formulation and practical implementation *Ocean Dynamics*. vol. 53, pp. 343-367

[7] Laghrouche S, Plestan F, and Glumineau A 2007 Higher order sliding mode control based on integral sliding mode *Automatica*. vol. 43 pp. 531-537

[8] Polyakov A and Poznyak A 2009 Reaching time estimation for "super-twisting" second order sliding mode controller via Lyapunov function designing *IEEE Transactions on Automatic Control*. vol. 54, pp. 1951-1955

IOP Conf. Series: Materials Science and Engineering **630** (2019) 012003 doi:10.1088/1757-899X/630/1/012003

The modelling of the crushing of meat raw materials for making management decisions in the production of emulsified meat products

B R Kapovsky, A N Zakharov and M A Nikitina

V.M. Gorbatov Federal Research Center for Food Systems of Russian Academy of Sciences, 26 Talalikhina Street, Moscow, 109316, Russia

b.kapovski@fncps.ru

Abstract. The article describes an innovative approach of the grinding heterogeneous frozen meat raw materials on the basis of single-stage milling method. In the technological cycle, in comparison with the traditional solution, the one shredder is used instead of three meat-cutting machines. The high quality of the finished product is achieved by modeling the grinding parameters. As a result of the analysis of the experimental data of the "new" technological process of meat grinding, an intelligent control system for the production of minced meat with machine quality control is proposed. Promising technology will reduce the cost of production of mass demand products that meet quality standards.

1. Introduction

Currently, almost all Russian enterprises of the meat processing industry according to various estimates (85-95)% are equipped with imported equipment [1-3]. In this regard, it is necessary to solve the problem of import substitution of expensive meat processing equipment in the light of the requirements of the Doctrine of the food security of the Russian Federation. The analysis of quality of the meat products developed by domestic producers, showed, that not all products can be estimated positively, and it is required to improve qualitative characteristics of the meat products made in the Russian Federation, at strict observance of normative requirements to technology of their production. The creation of energy-and resource-saving technology for the production of meat products while reducing their cost and guaranteed quality on the automatic line, equipped with types of technological equipment of Russian production, will help to solve these problems. The theoretical aspects of the processing of frozen raw meat cutters are developed in V.M. Gorbatov Federal Research Center for Food Systems of Russian Academy of Sciences: the technique of analytical calculation of the performance of the cutting process meat cutter with helical teeth, the relative error not exceeding 6%; the calculation method of energy consumption for the cutting process of the raw materials by the method of milling on the basis of application specific cutting forces; the system of automatic process control of minced meat production with the use of the control computer is proposed, which provides a guaranteed high quality of the finished product [4, 5]. From the raw materials obtained at the current experimental installation of one-stage grinding of frozen meat by the method of milling, samples of finished products (cooked sausages, semi-finished products, canned food for kids) were developed, which surpass the similar products produced by traditional technology in terms of quality characteristics. Currently on the basis of the one-stage crushing of meat raw materials the automatic line on development of meat products of food with control system of quality of minced meat on the basis of artificial intelligence on the principle of "deserted technology" is created. As a result of a significant reduction in the cost of finished products, energy saving and improving sanitary and hygienic indicators in its production, the meat products obtained by the proposed technology will be available to the mass buyer, including the poor.

2. Research tasks

The structural and mechanical properties are determined largely by the size of the particles of the crushed raw material, noted in [6-9], means, the degree of its grinding. It was proved [8-12], that with the help of structural and mechanical properties it is possible to control the technological parameters of raw materials and minced meat, the product quality at any stage of the technological process of minced meat preparation, as well as the consistency of finished products. It follows from this, that the machine control of the degree of grinding of the initial frozen meat raw materials largely determines the control of the entire technological process on the developed automatic line for the production of emulsified meat products. Taking this into account, it is possible to formulate the objectives of the management of a single-stage process of grinding frozen meat raw materials by milling method. Firstly, to maintain a given degree of grinding during heterogeneity of raw meat on the structural and texture characteristics. Secondly, to calculate the means of automatic control system (ACS) technological process forecast changes in the size of meat particles (meat chips) in the grinding process. In accordance with these objectives should solve the problem of synthesis of ACS. For this need: to choose the structure of the system, its elements and the topology of cause-and-effect relationships between them; to develop algorithms for control devices and the values of their parameters, for example, the settings of regulators [13]. During the cutting of frozen meat by cutters have a significant difference. The quality of grinding is significantly affected by the heterogeneity of frozen meat raw materials on the structural and texture characteristics. Different amounts of ice (water) in frozen meat at different storage temperatures; the presence of fat and connective tissue are structural signs of heterogeneity. Different orientation of the muscle fibers in the unit volume of the product relative to the blades of the cutter is a textural indication of the heterogeneity of raw materials. An approximate picture of the heterogeneity of the block of frozen meat on the structural and texture characteristics in the orientation of the cutting surface of the block relative to the cutter is shown in Fig. 1.

Figure 1. Heterogeneity of frozen meat block.

Trace of the cutting edge of the cutter with a screw tooth (Fig. 1) will pass through the fibers of muscle tissue, differently oriented with respect to it (longitudinal and transverse), through other types of tissues, volumes of ice of different shapes, sizes and orientations relative to the cutting edge. When placing the longitudinal beam of muscle fibers relative to the cutting blade of the cutter, the width of the cut layer of meat will be greater in comparison with the transverse arrangement of the same fibers. The impact of the above factors influencing the degree of grinding is random, so the statistical analysis of the size of the meat chips, chopped meat, and energy for cutting frozen meat raw materials of the multiblade tool.

3. Materials and methods of research

It was created the experimental set-IBF-1 (chopper units milling modification 1) for study the one-step process of cutting a block of frozen meat cutters. The design of the IBF-1 is schematically shown in Fig. 2.

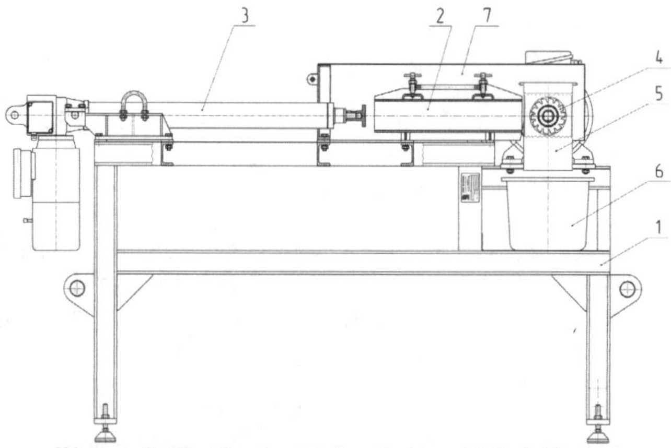

Figure 2. The device of the chopper IBF-1 (figure)
1 - frame; 2 – guide chute; 3 – feed mechanism; 4 – cutting mechanism; 5 – casing; 6 – prefabricated container; 7 – cover; 8 – control cabinet (not shown in the figure).

The installation was completed with cutters of different design and geometry. The parameters of the cutting mode (the speed of the cutter and the feed rate in the grinding zone) were set by the corresponding settings of the frequency converters of the supply voltage working on the electric motors of the cutting mechanisms and the supply of the IBF-1 installation. The experimental blocks of frozen meat, cut from blocks of industrial size (bovine veneered second grade and pork veneered semi-fat), were ground at the plant. According to [14], the mass fraction of connective and adipose tissue in beef was no more than 20%, in pork the mass fraction of adipose tissue was in the range from 30% to 50%. The temperature of the meat in the center of the experimental block before grinding is -12° C-14° C. The placement of the experimental frozen meat block in the guide chute of the IBF-1 chopper is shown in Fig. 3.

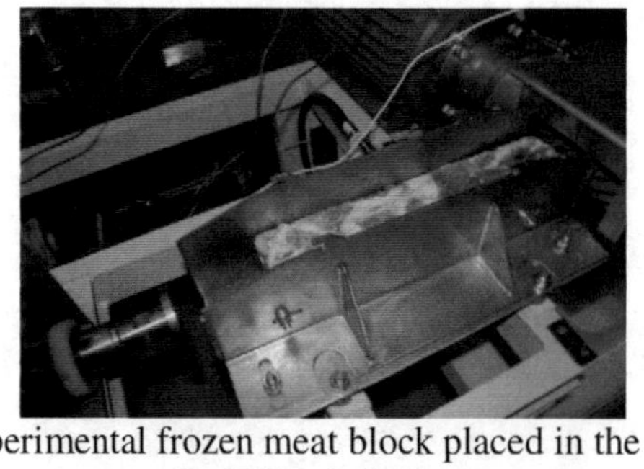

Figure 3. Experimental frozen meat block placed in the guide chute of the IBF – 1 chopper.

The movement of the blocks to the mill was carried out by the impact on them of the rod of the chopper feed mechanism with a given feed rate on the guiding planes. The guiding planes of the working chamber IBF-1 were exposed and fixed in accordance with the size of the experimental blocks of frozen meat. The ground meat was extracted from the combined container and subjected to microstructural study. Microstructural studies were carried out in accordance with the current regulatory documentation [15]. Histological preparations were made on the freezing microtome MICROM HM 525. The sections were studied using a light microscope "Axio Imager.A1" (Carl Zeiss Germany) with the using of the computer program for image analysis.

The active power consumed by the electric drive of the cutting mechanism of the installation in the operating mode was measured and recorded by the industrial analyzer-registrar ASM-3192. Statistical analysis of the parameters of the grinding process (particle size of crushed meat – meat chips, and energy consumption for cutting frozen meat cutter) was carried out by calculating the numerical characteristics (mathematical expectation and dispersion) of the experimental distributions of these parameters obtained during the grinding of raw materials at the IBF-1. Verification of compliance of the law of experimental distribution of parameters as random variables with the proposed hypothetical distribution law was carried out according to Pearson's consent criterion [16, 17].

4. Research results and discussion

The linear dimensions of the meat chips (thickness and width) of the crushed meat obtained at the IBF-1 unit grinding of the initial frozen raw material (beef) are shown in Fig. 4. The data are obtained as a result of microstructural analysis of the crushed raw material sample.

Figure 4. The size of the meat chips (beef)
a_i –- chip thickness, mcm; b_i –- chip width, mcm.

The histogram of the density distribution of the frequency of ingress of the thickness of the meat chips, corresponding to Fig. 4, in the range of its values is shown in figure 5.

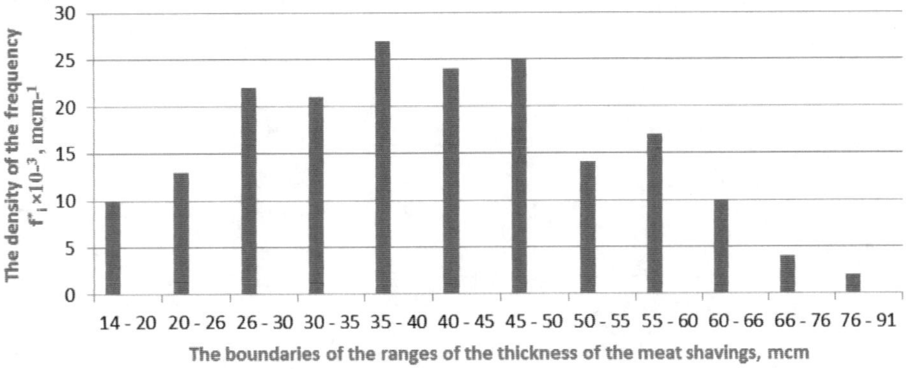

Figure 5. Histogram of the distribution of the frequency density f_i^* ingress of the thickness of the meat chips in the range of its values (beef).

Solving the problem of equalizing the obtained statistical distribution, we hypothesized that the test (experimental) distribution corresponds to the normal law. The likelihood of this hypothesis was established by methods of mathematical statistics. Similarly, the experimental active power distribution was proved to be consistent (Fig. 6), consumed by the electric drive of the chopper cutting mechanism IBF-1 in operating mode, the law of normal distribution (Gauss).

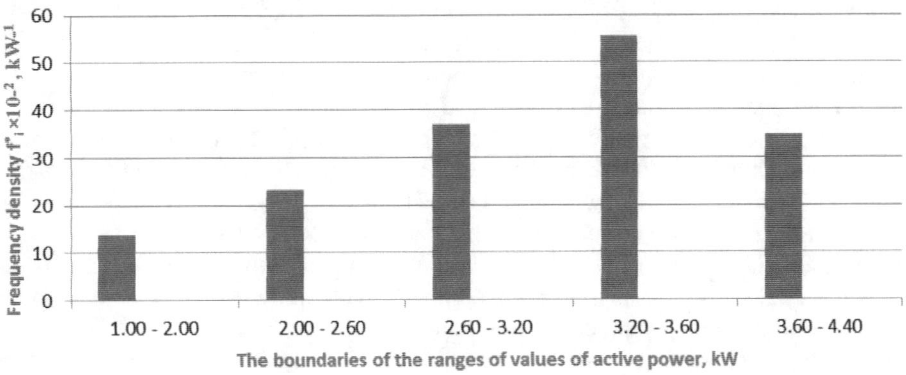

Figure 6. Histogram of the power distribution frequency density consumed by the drive motor of the cutting mechanism of the IBF-1 installation in the operating mode, by the ranges of its measured values.

From the results of statistical analysis of the experimental distribution of active power consumed by the drive motor of the cutting mechanism of the IBF-1 installation in the grinding process, it follows that the average square deviation of the power is $\approx$ 26% of the value of the expectation estimate. In this regard, it can be concluded that the moment of resistance to grinding, that is, the load on the mill in the operating mode, varies significantly in magnitude. This is due to the significant heterogeneity of the feedstock, as noted above. When changing the load on the mill grinder in the operating mode, the frequency of its rotation changes, which leads to additional dispersion of the linear dimensions of the meat chips due to changes in the parameters of the cutting mode (the feed to the tooth of the cutter and the cutting speed of raw materials). This factor of influence on the degree of grinding of meat reduces the homogeneity of the produced minced meat, which leads to a decrease in the quality of meat products. Taking this into account, it is necessary to ensure the stabilization of the parameters of the cutting mode by means of the automatic control system of the grinding process in the operating mode of the grinder. The application of automatic control of one-stage technological process of minced meat production is shown by the example of the developed line for the production of emulsified meat products (Fig. 7).

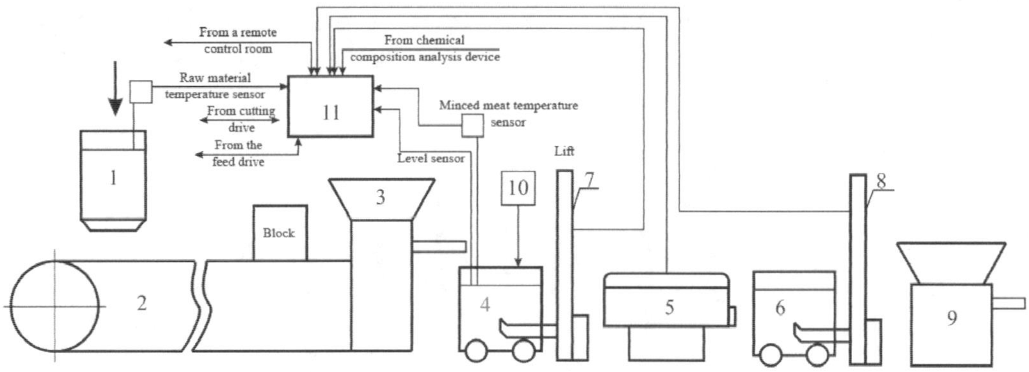

Figure 7. Automatic sausage production line

1 - input control equipment of raw materials; 2 – conveyor; 3 – milling shredder; 4 – storage hopper dispenser; 5 – minced meat; 6 – truck; 7, 8 – lifts; 9 – filling machine (syringe); 10 – device analysis of the chemical composition of the crushed meat; 11 – control computer (CC).

During the implementation of this project, the task was not only to improve the technological process of cutting frozen meat raw materials through the use of a new method of grinding by milling, but also to ensure the production of minced meat of a given quality on the automatic line.

For solving this problem the line was equipped with storage bins-dispensers, in which the required amount of crushed raw materials of each type was measured in accordance with the recipe of the finished product. The line is also equipped with equipment for monitoring the chemical composition of crushed meat (protein, fat, moisture) in the real time stream. In the process of mixing the minced meat in the minced meat mixer, the viscosity is controlled by a sensor and the pH of the minced meat is measured, water and other components provided by the recipe are dosed. These measurements are fed into an industrial computer that controls the technological process of producing emulsified meat products on an automatic line. For the synthesis of ACS technological process on the proposed line, in particular the problem of modeling the size of the meat chips in the grinding process, it is necessary to take into account the impact on the mill from the load. The above data give reason to assume in the first approximation the normal law of load distribution on the mill when grinding blocks of raw materials of industrial sizes. The analysis of experimental data allows us to assume a random process of loading on the mill stationary process, which has the property of ergodicity. This means that any implementation of a stationary random process of a certain duration sufficiently fully represents the entire set of realities of the process under consideration. Taking into account the results of experimental research, for mathematical modeling of one-stage process of grinding frozen meat raw materials in order to predict the size of the resulting meat chips, it is advisable to use the method of forming filter (FF) [17] to simulate the perturbation acting on the system from the load (Mc in Fig. 8).

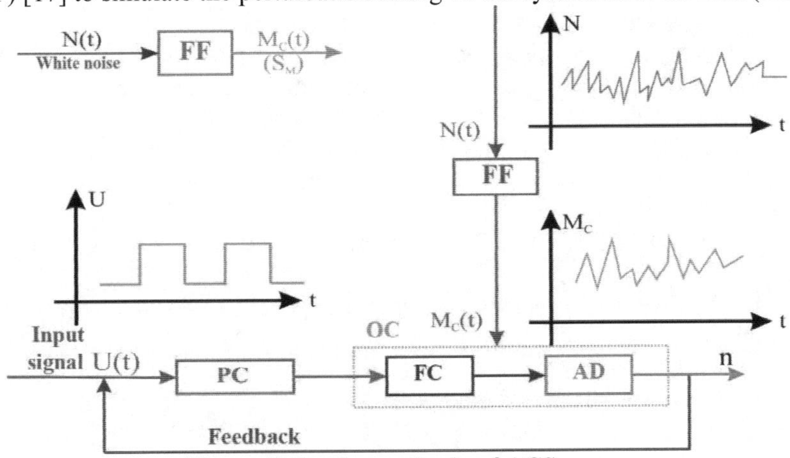

Figure 8. To the synthesis of ACS

OC – the object of control; FC – frequency converter that changes the speed of the cutter; AD – asynchronous motor drive mechanism of cutting the grinder; N(t) – white noise; U(t) – input signal (setting the cutting speed of raw materials); FF – forming filter; PC – speed controller (frequency) of rotation of the cutter; n – the speed of the cutter.

Control of one-stage grinding provides for the maintenance in automatic mode of the established degree of grinding of raw materials with the stabilization of the parameters of the cutting mode in real time. ACS technological process on the developed automatic line for the production of emulsified meat products has the signs of an intelligent control system [19]:

1) the presence of close information interaction of the intelligent system with the environment when using information communication channels - in case of control of one-stage grinding of meat blocks is controlled by their temperature, the presence of foreign matter, the weight of the blocks;

2) the existence of simulation predictions of changes in the external environment and their own behavior - in this case, the system simulates the size of the meat chips, obtained by cutting raw materials multiblade tool;

3) increasing intelligence and improving their own behavior - the system is trained in real time, specifying the forecast size of the meat chips in the process.

The algorithm of the ACS is characterized by the fact that, in addition to the statistical information obtained during the grinding of a real block of meat about the change in the speed of the cutting shaft under the influence of the moment of resistance to grinding, the industrial computer will have an additional amount of similar statistical information in the computer simulation of grinding "virtual" blocks of meat. This will calculate the point and interval estimates of the change in the frequency of rotation of the cutting shaft of the grinder in the grinding process, and on their basis, according to the established analytical dependence, to determine the same estimates for changing the characteristic size of the resulting meat chips with a given statistical accuracy and reliability. Besides, dispersions of the estimates themselves are necessarily calculated, that is, determines the degree of "blurring" of the boundaries of the range in which the size of the meat chips is located, which is important in the production of emulsified meat products. In the process of real grinding of block meat in the industrial computer accumulates a database and knowledge about the process and parameters of grinding, that is, the system is trained in the operating mode, improving its work on predicting the degree of grinding of raw materials. As a result of the implementation of the above algorithm, the control system will have statistical information about the degree of grinding of raw materials in digital form. The structure of the ACS may include equipment for rapid analysis of the chemical composition of the crushed meat, as well as its temperature after grinding. Then the further process of minced meat production can be controlled strictly in the function of time, excluding the subjective factor of assessing the degree of readiness of the final product. This will allow you to get the finished product of a given high quality.

5. Conclusion

Thus, an innovative approach to obtaining minced meat of a given quality, while strictly controlling the degree of grinding and readiness of the final product. The organization of the quality control system of minced meat on the equipment does not require complex technical solutions, and is achieved by using software with simulation of grinding parameters. At the same time, the control system operates in the mode of interaction with the external environment, is self-learning in the process of work and predicts the influence of external factors on the process of grinding meat raw materials

6. References

[1] Maksimov D A and Zhukov V G 2013 Modern equipment for grinding frozen meat blocks *Meat industry* no. 2 pp 31-33
[2] Solov'ev O V 2010 *Meat processing equipment of a new generation. Directory* (Moscow: DeLy print) p 467
[3] Ivashov V I 2007 *Technological equipment of enterprises of the meat industry. Part II. Meat processing equipment* (Saint Petersburg: GIORD) p 464
[4] Lisitsyn A B, Ivashov V I, Zakharov A N, Kapovsky B R and Maksimov D A 2013 Chopping block of frozen meat by milling *Vsyo o myase* no. 4 pp 42-46
[5] Lisitsyn A B, Ivashov V I, Zakharov A N, Kapovsky B R and Kozhevnikova O E 2013 Intelligent control system of quality minced meat *Vsyo o myase* no. 6 pp 32-36
[6] Gorbatov A V 1968 *Rheology in the meat industry* (Moscow: CRIITER) p 67
[7] Machikhin S A, Maslov A M, Tabachnikov V P, Machihin Yu A and Kosoy V D 1982 *Structural and mechanical characteristics of food products. Directory* (Moscow: Light industry) p 296
[8] Ilyuhin V V 1990 *Physical-technical fundamentals of cryo divisions of food* (Moscow: Agropromizdat) p 207
[9] Kosoy V D, Malyshev A D and Yudina S B 2005 *Engineering rheology in the production of sausages* (Moscow: ColosS) p 264
[10] Alekseev E L. and Pakhomov V F 1987 *Modeling and optimization of technological processes in the food industry* (Moscow: Agropromizdat) p 325
[11] Ivashkin Yu A, Protopopov I I, Borodin A V, Kopchikov A V and Shutov S A 1987 *Modeling of production processes of meat and dairy industry* (Moscow: Agropromizdat) p 232
[12] Pestretsov S I 2009 *Computer modeling and optimization of cutting processes: textbook* (Tambov: publishing house of TSTU) p 104
[13] Pupkov K A, Egupov N D and Voronov E M 2004 *Methods of classical and modern theory of automatic control* (Moscow: publishing house of Bauman Moscow State Technical University) chapter 3, p 616
[14] GOST R 54704-2011 *Frozen blocks of venated meat. General specifications* (In Russian)
[15] GOST 31479-2012 *Meat and meat products. Method of histological identification of the composition* (In Russian)
[16] Chernoff H and Lehmann E L 1954 The use of maximum likelihood estimates in $\chi 2$ test for goodness of fit *The Annals of Mathematical Statistics* Vol 25 pp 579-586
[17] Greenwood P E and Nikulin M S 1996 *A guide to chi-squared testing* (New York: John Wiley & Sons) p 280
[18] Pupkov K A, Egupov N D and Voronov E M 2004 Methods of classical and modern theory of automatic control (Moscow: publishing house of Bauman Moscow State Technical University) chapter 2, p 640
[19] Pupkov K A, Egupov N D and Voronov E M 2004 Methods of classical and modern theory of automatic control (Moscow: publishing house of Bauman Moscow State Technical University) chapter 5, p 784

Model of system dynamics to study the effects of unmanned vehicles introduction in the economic environment

Alla Brom, Kirill Belonosov and Olga Vereschagina

Bauman Moscow State Technical University, Moscow, Russian Federation, 105005

abrom@yandex.ru

Abstract. The article is devoted to the study of problems appearing from the introduction of unmanned vehicles in the economic environment. The authors propose an approach to visualize and study the mutual influence of factors causing the emergence of such problems, based on cause-effect diagrams of system dynamics. The main factors investigated are: the investment advisability, the activity of trade unions and the main components of the sustainable development concept. The interaction of the described factors was modeled using Anylogic environment flowcharts, the results were obtained. The study showed that the transition to unmanned vehicles will reduce the costs of transport companies and the intensity of emissions on the background of social conflict and increasing drivers union activity.

1. Introduction

Nowadays, unmanned vehicles of companies such as Tesla and Google, are already actively used in the US and EU countries. However, with the introduction of unmanned vehicles in the real economic environment, appears a large number of specific factors that are important not only by themselves, but also interact with each other. [1-3]. This problem is given much attention in different countries of the world. [4]. There is a need to study and visualize this interaction, as it plays the greatest role in making management decisions regarding the development of approaches to the introduction of unmanned vehicles.

2. Research Method

All complex systems include variables that interact with each other through positive and negative feedback loops. As a means of visualizing the variables interaction, it is proposed to use cause-effect diagrams used in system dynamics [5, 6]. The diagram serves as a visual aid on the interaction and role of variables. In case of the positive relationship, increasing / decreasing the first element contributes to increasing / decreasing the second element. The results that can be obtained from cause-effect diagrams are limited by their qualitative, conceptual role. [7]. However, the diagram allows to conclude about the functioning principle of the system, which is very useful in studying the modern digital technologies influence on the environment.

3. Results and discussion

Obviously, each new unmanned vehicle in the company's park will be equivalent to one dismissed driver. Increasing the number of layoffs leads to an increase in driver dissatisfaction in the company, which may influence management decisions through a trade union organization inside the enterprise, strikes, absenteeism and other leverage. Thus, a conflict arises: on the one hand, replacing a traditional fleet with an unmanned vehicle leads to fuel consumption reduction. But on the other hand, a huge number of people involved in the motor vehicles management will remain without work. [6]. The investment advisability increases depending on the number of new cars, and the whole process will have an exponential growth pattern. Therefore, a balancing feedback loop should be present in the dynamic model, preventing the exponential growth. The result is an increase in the discontent degree of dismissed drivers and an increase in the activity of trade union organizations. The consequence is an incentive for strikes and other levers of influence on the management of the enterprise. (Figure 1).

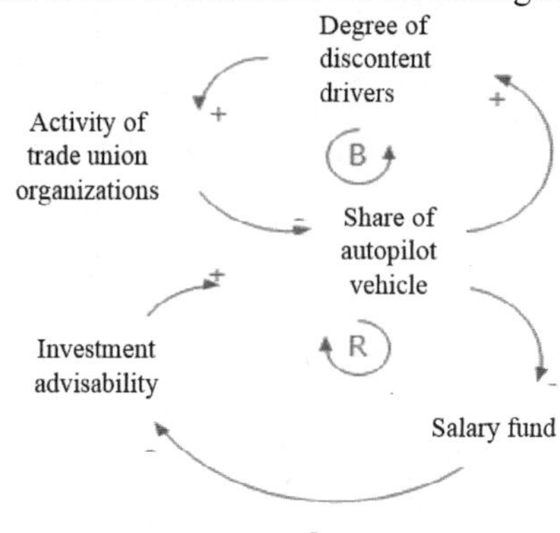

Figure1. The interaction of the balancing and amplifying cycles.

The process of replacing the fleet will also form a balancing feedback loop, due to the processes occurring in the organization. Its formation can be explained by the following logic - with each new driver dismissed, a growing degree of discontent in the society of drivers will take place, and trade union organizations will have an ever-increasing impact on the company's management, preventing the replacement of the fleet. Reduced fuel consumption leads to lower exhaust emissions. (Figure 2).

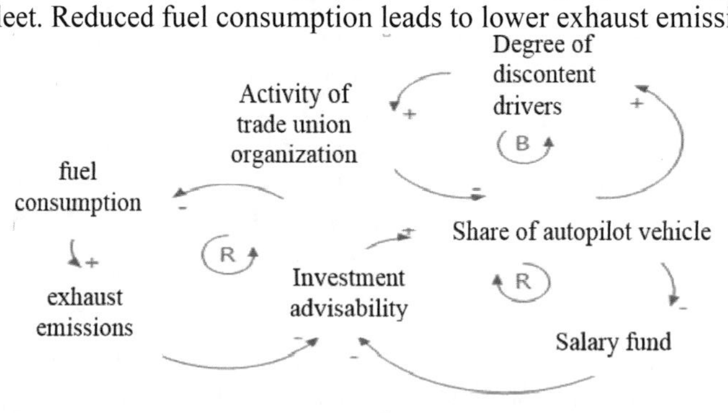

Figure 2. Cause-effect diagram illustrating the reduction of fuel consumption effect

Transportation costs decrease as salary fund decreases. This will reduce the price of the finished product and thus allow to occupy a larger market share. Increasing the number of customers forms the economies of scale effect, which again reduces the price and makes possible to take even greater market share. This cycle is a company growth accelerator due to the introduction of autopilot systems. The insurance cost depends on the frequency (probability) of getting into a traffic accident. Thus, companies should expect a reduction in vehicle fleet insurance costs. [8].

In that way the positive aspects of digital technologies in transport will be weakened by rising activity of drivers trade unions, which can eventually lead to social conflicts in society. The drivers have the ability to influence management decisions, trade union organizations, strikes, absenteeism and other leverage. In case if most of the cars function autonomously, then with much greater probability the above losses will be less.

Table 1. Elements used in the flow diagram.

Element name	Symbol	unit of measurement
Drivers on staff	*V*	person
Dismissed drivers	*C*	person
Rate of dismissal	*U*	person / month.
Investment advisability	*inv*	transport units
Salary fund	*Fond*	ruble
Novelty factor	*K*	-
Average salary of drivers	Z_{cp}	ruble
Number of cars in the park	$K_{парк}$	Units of transport

Staff and dismissed drivers were taken as accumulators, between which there is a discharge rate. (Figure 3):

$$C(t)/dt = U \quad (1)$$

$$V(t)/dt = -U \quad (2)$$

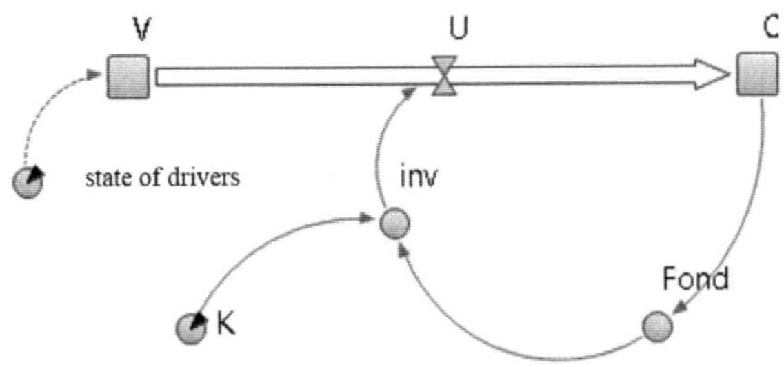

Figure 3. Intermediate flowchart creation stage

We introduce the coefficient of novelty K. The enterprise management has the right to decide for itself how many cars to replace at the initial stage depending on financial possibilities and the strategy being implemented: this can be either one car or the entire fleet at once. [9]. Thus, this parameter adds flexibility to the model being developed:

$$1 < K < K_{парк} \quad (3)$$

Obviously, the feasibility of investing a project that does not bring any benefit is 0. In the object under study, the dismissal of employees contributes to saving money and, consequently, increasing the company's revenue to *Zcp* per month. In order to determine the benefits of investing, we must also have data on costs. Let us assume that the project's profitability is high and the implementation of the technology is based on the project implementation period minimizing criteria. Then the process of introducing cost-effective investments will be a snowball, in which the volume of current investments depends on the previous value. Therefore, we assume that the replacement of one car increases the feasibility of investing in one car more:

$$Inv = K + Fond/ Zcp \quad (4)$$

Thus, there is all the necessary data for the compilation of the subtotal. In our example, the following parameters: model time: months; staff of drivers: 100 people. According to the simulation results, with given parameters, after 3.5 months all drivers in the enterprise will be dismissed and the monthly salary savings will be equal to the value of the variable *Fond*, as a result of modeling that is 5 300 000 rubles.

We believe that the dismissed drivers can influence employees on staff. We introduce some additional parameters presented in Table 2.

Table 2. Additional elements of the model

Element name	Symbol	unit of measurement
Degree of discontent drivers	*N*	-
The average number of people in the immediate vicinity of one driver	*O*	person
Power of influence	*F*	-
Union activity	*A*	person
Fuel economy	*FE*	ruble
The probability of finding a driver in the company	α	%
Emission reduction	*Gas*	

For a general understanding of the picture we will reveal the significance of some elements: *N* - degree of drivers discontent. This variable has two parameters that characterize the degree of discontent: *O* – communication. The average number of people with whom one driver communicates. Let the driver communicate with 5 people on average: *O*=5. *F* – power of influence. Parameter taking value in the interval $1/1000 < F < 1$. Assume that the value of this parameter is 1/500:

$$F = 1/500 \tag{5}$$

Thus, from the accumulator C "dismissed drivers" there is a positive connection to the variable *N*, which depends on communication and the strength of influence. It is necessary to add a variable that refletcs degree of driver discontent and the likelihood that the dismissed driver does not have a repeated effect.

$$\alpha = V/100 \tag{6}$$

$$N = C \cdot \alpha \cdot O \cdot F \tag{7}$$

In that way *N* takes a value from 0 to 1. At the maximum value of this variable, the activity of the trade unions becomes the greatest. Therefore, it is true that the more displaced drivers a transport company has, the more they influence people who have access to leverage. In this regard, it is necessary to enter the parameter of trade unions activity *A*, denoting the number of people who cannot be dismissed at a given time. After part of the drivers dismissed, social tensions increase in the work team. Therefore, the more drivers dismissed, the greater the degree of discontent is, but as the number of drivers in the company becomes smaller, they can not be so active as before:

$$A = V \cdot N \tag{8}$$

Suppose autopilot trucks use less fuel by 6%. Based on this, we introduce a new variable *FE* - fuel savings that will have one parameter *p1* – current fuel consumption of 15 liters per 100 km.

$$FE = 0{,}06 \cdot p1 \cdot C \tag{9}$$

Dynamic variable *FE* will serve as a source of information on the amount of fuel saved. An important factor has been and remains the reduction of exhaust emissions due to reduced energy consumption. To determine a quantitative estimate of emission reductions, we introduce a new variable *Gas* – CO_2 reduction and *p2*- conversion rate. In this way:

$$Gas = FE \cdot p2 \tag{10}$$

The result of reducing exhaust emissions is the amount of fuel saved multiplied by the conversion factor.
Now lets model the created flowchart in the Anylogic environment. (Figure 4).

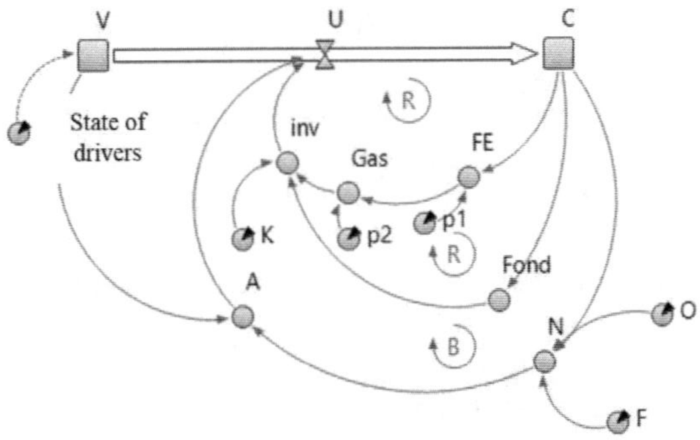

Figure 4. Streaming flowchart

Table 3. A general summary of the model's elements and their formulas

Element	Element type	Formula or value	unit of measurement
V	accumulator	*V(t)/dt = -U*	person
C	accumulator	*C(t)/dt = U*	person
U	flow	*inv = - A*	person
Fond	dynamic variable	*Fond = -C · Zcp*	ruble
Zcp	parameter	*53000*	ruble
inv	dynamic variable	*inv = K + Fond/ Zcp*	person
K	parameter	*3*	person
N	dynamic variable	$N = C \cdot \alpha \cdot O \cdot F$	-
α	parameter	$\alpha = V/100$	%
O	parameter	*5*	person
F	parameter	*1/500*	1/person
A	dynamic variable	$A = V \cdot N$	person

Lets list the simulation results of the interaction between the contours of the balancing and amplifying cycles in Table 4.

Table 4. The simulation results of the interaction between the contours of the balancing and amplifying cycles

Symbol	Result	unit of measurement
Ткон	8.0806	months
CC	100	person
V	0	person
Fond	5 300 000	ruble
N	1	-
U	100	person/month
A	0	person

The simulation results showed that with an increase in the activity of trade unions, dismissal takes place over a long period, during which the effect of the balancing loop is observed. At a certain point in time, a crucial situation occurs when the activity of trade unions begins to fall sharply. During this period, there is a rapid increase in the number of layoffs. This is due to the drivers loss of leverage on the company. Considering the transport infrastructure of digital economy, it is important to take into account the flow of processes in dynamics, observing the mutual influence of factors that will inevitably accompany the introduction of automation innovations in practice. [10,11].

4. Conclusion

The results of the study showed that the management of transport, trade, logistics companies and the government of the country should be ready for social conflicts in society related to the implementation

of automated systems. We should highlight that different regions and climatic zones have a significant impact on the automation growth rate. The influence of weather conditions can be taken into account by introducing the corresponding dynamic variable into the model.
In 2019, six organizations are planning to begin trial operation of highly automated cars in Russia - PJSC KamAZ, Avrora Design Bureau LLC, Scientific-Design Bureau of Computing Systems, University Innopolis, Moscow Automobile and Road State Technical University (MADI) and Yandex. In this regard, we can conclude that the validation of the model will be possible in the near future. However, today the automation technologies development is far ahead of the development of legislation in this sphere. Therefore, the speed and success of the autopilot vehicles introduction will largely depend on the creation of an appropriate legislative framework, which should be created as soon as possible.

References

[1] A.E. Brom, K.U. Belonosov, Study of the problems of introducing unmanned vehicles into the economic environment. Bulletin of MGOU. Series: Economy. №1, (2018).
[2] A.G. Yuzaeva, V.V. Kukartsev, Unmanned vehicles: dangers and development prospects. Actual problems of aviation and astronautics. №12, (2016).
[3] J. Bierstedt, A. Gooze, C. Gray, J. Peterman, L. Raykin, and J. Walters, Effects of Next-Generation Vehicles on Travel Demand and Highway Capacity. FP Think, (2014).
[4] R. Hoogendoorn, B. Van Arem, and S. Hoogendoorn, Automated driving, traffic flow efficiency, and human factors. Transportation Research Record. vol. 2422, pp. 113–120, (2014).
[5] J. Gharajedaghi, Systems Thinking. Managing Chaos and Complexity: A Platform for Designing Business Architecture. Minsk, Grevtsov Books Publ., 480 p, (2010).
[6] V.A. Shakhnov, L.A. Zinchenko, V.V. Kazakov, V.V. Makarchuk, E.V. Rezchikova, Cognitive visualization of carbon nanotubes structures, Advances in Intelligent Systems and Computing 874, pp. 273-279 (2019).
[7] N.N. Barbashov, I.V. Leonov, K.D. Sologub, Modeling of dynamic and economical characteristics of life-saving device with flywheel energy storage, Lecture Notes in Mechanical Engineering (9783319956299), pp. 151-158 (2019).
[8] M.M. Morando, T. Qingyun, L.T. Truong, Hai L. Vu, Studying the Safety Impact of Autonomous Vehicles Using Simulation-Based Surrogate Safety Measures. Journal of Advanced Transportation. Volume 2018, 11 pages, (2018).
[9] A.A. Sukhobokov, R.Z. Galimov, A.A. Zolotov, A Strategic Management System Based on Systemic Learning Algorithm, Advances in Intelligent Systems and Computing 848, pp. 290-295 (2019).
[10] E. Lyapuntsova, Y. Belozerova, , I. Drozdova, G. Afanas'Ev, E. Okunkova, Entrepreneurial Risks in the Realities of the Digital Economy, MATEC Web of Conferences, 251,06032, (2018).
[11] I. Alpackaya, D. Alpackiy, Perspectives and consequences of implementation and development digital economy, MATEC Web of Conferences 193, 05087, (2018).

IOP Conf. Series: Materials Science and Engineering **630** (2019) 012005 doi:10.1088/1757-899X/630/1/012005

Analysis of the dynamics of movement landing vehicle with an inflatable braking device on final stage of trajectory

Vsevolod V. Koryanov[1,a], Victor P. Kazakovtsev[1], Helene Griselin[2] and Chen Danhe[3]

[1]Bauman Moscow State Technical University, 105005, 2-nd Baumanskaya, 5, b1, Moscow, Russian Federation
[2]École nationale supérieure d'arts et métiers, Paris 151 Blvd de l'Hôpital, F-75013, Paris France
[3]NJUST School of Mechanical Engineering, Nanjing, 200 Xiaolingwei Street, Xuanwu District, Nanjing, Jiangsu, 210094, China

[a]vkoryanov@mail.ru

Abstract. For various missions that are currently being designed and planned at the present time, it is necessary to ensure reliably the stage of landing on the surface of the planet. Among the currently proposed devices that help to perform the landing of the landing vehicle, a separate interest is inflatable devices. Using inflatable device instead of rigid device has advantages, such as a small mass and a great compacity. Studies have focused on the angular motion of landing vehicles entering the atmosphere, with allowance of small asymmetries. During the descent in the atmosphere, the landing vehicle and its braking device are subjects to important perturbations, due to external environment. Thus, the inflatable device can be deformed, due to its flexibility, and the movement can become unstable. In the paper is analyzes the movement of such a landing vehicle, which uses such special devices made of inflatable devices. The paper is devoted to the analysis of movement at the final stage of landing - approaching the surface, touching the surface and further interaction with the surface. As a result, it is concluded that the most unfavorable approach to the surface, when the resulting overload is of the greatest importance.

1. Introduction

Currently, a large number of space missions are being designed. It is increasingly important to carry out missions that provide for the return of cargo from an orbit or space to small automatic stations. This may be the return of samples in interplanetary missions, there may also be the return of cargo from orbit around the Earth and so on. For such missions, it is necessary to perform a reliable landing of the landing vehicle. To do this, it is necessary to perform braking in the atmosphere at very high speeds when entering the atmosphere and to acceptable speed values when approaching the surface [1]. It is possible to use various devices that help to perform such braking [2-5].

This paper is devoted to the final stage of landing, when contact with the surface occurs, large values of overload occur and it is necessary to analyze the search for optimal variants of the initial parameters of the approach to the surface (Figure 1, stage IV).

2. Theoretical part: determination of the angular motion of landing vehicles entering atmosphere

One of the interesting and effective options is the use of inflatable devices to perform a successful landing. Such devices can be used for the entire length of the landing - from the moment of entry to the moment of contact. An example of such a project is the MetNet project [6, 7], which is well described in scientific works and papers [8, 2] (Figure 1).

Previously, the authors considered various stages of the landing vehicle in the atmosphere of the planet, to which articles are devoted [2]. Materials of other scientists who work in the field of flexible structures are presented here [9, 10]. In addition, there are a number of articles on deployable structures in space that are also of interest to our research [11-13].

In this part, we will study the angular motion of a landing vehicle in the atmosphere with allowance of small asymmetries [2]. To brake into the atmosphere, we will consider that the landing vehicle has inflatable braking device, and that there are two of them: the primary braking device, used in the upper atmosphere, and the additional one, used in the lower atmosphere. During the descent, the landing vehicle and its braking device are subjects to important perturbations, due to external environment. Thus, the inflatable device can be deformed, due to his flexibility, and the movement can become unstable.

Let's determine the equations of the angular motion. We will use a method that analyze how the design parameters and the aerodynamics coefficients influence the asymmetry of a deviation from the longitudinal axis of the vehicle velocity vector and give them coefficients corresponding to the degree of their influence. We suppose that there is a rapid development of resonance modes of motion.

We can write:

$$\omega_x = \frac{d\omega_x}{dt} = \frac{1}{J_x}\left[J_{xy}\cdot(\omega_y-\omega_x\cdot\omega_z)+J_{xz}\cdot(\omega_z+\omega_x\cdot\omega_y)+J_{yz}\cdot(\omega_y^2-\omega_z^2)+(J_y-J_z)\cdot\omega_y\cdot\omega_z+q\cdot S\cdot l\cdot\left(m_{x_0}+m_x^{\bar{\omega}_x}\cdot\frac{\omega_x\cdot l}{V}++C_y\cdot\frac{\Delta z}{l}-C_z\cdot\frac{\Delta y}{l}\right)\right]$$

$$\omega_y = \frac{d\omega_y}{dt} = \frac{1}{J_y}\left[J_{xy}\cdot(\omega_x+\omega_y\cdot\omega_z)+J_{xz}\cdot(\omega_z^2-\omega_x^2)+(J_z-J_x)\cdot\omega_x\cdot\omega_z+J_{yz}\cdot(\omega_z-\omega_x\cdot\omega_y)+q\cdot S\cdot l\cdot\left(m_{y_0}+m_y^{\beta}\beta+m_y^{\bar{\omega}_y}\cdot\frac{\omega_y\cdot l}{V}+C_x\cdot\frac{\Delta z}{l}\right)\right] \quad (1)$$

$$\omega_z = \frac{d\omega_z}{dt} = \frac{1}{J_z}\left[J_{xy}\cdot(\omega_x^2-\omega_y^2)+J_{xz}\cdot(\omega_x-\omega_y\cdot\omega_z)+J_{yz}\cdot(\omega_y+\omega_x\cdot\omega_z)+(J_x-J_y)\cdot\omega_x\cdot\omega_y+q\cdot S\cdot l\cdot\left(m_{z_0}+m_z^{\alpha}\alpha+m_z^{\bar{\omega}_z}\cdot\frac{\omega_z\cdot l}{V}-C_x\cdot\frac{\Delta y}{l}\right)\right].$$

Where: ω_x, ω_y, ω_z are the projections of the angular velocity of the landing vehicle on the axis of the related coordinate system;

$q = \frac{\rho V^2}{2}$ is the dynamic pressure;

V is the velocity of the landing vehicle;

S is the mid-section area;

l is the length of the landing vehicle;

J_x, J_y, J_z are the principal moments of inertia;

C_x, C_y, C_z are the aerodynamic coefficients of axial and transverse forces.

m_{x0} is the aerodynamic moment coefficient relative to the longitudinal axis;

$m_y^{\beta}, m_z^{\alpha}$ is derivatives of aerodynamic coefficients on the angle of attack (α) and glide (β);

$m_x^{\varpi_x}, m_y^{\varpi_y}, m_z^{\varpi_z}$ are the derivatives of the aerodynamics coefficients on the projections of the angular velocity of the landing vehicle on the axis of the related coordinate system;

Δy, Δz are the lateral displacements of the real center of mass relative too the longitudinal axis of the landing vehicle;

J_{xy}, J_{yz}, J_{xz} are the centrifugal moments of inertia;

m_{y0}, m_{z0} are the aerodynamic coefficients of transverse moments at zero values of angles of attack and glide angle

Several studies show that the duration of the resonance regime is around several seconds. We can then, in our equations above, neglect the influence of gravity in the dynamic equations of translational

motion, and the products of angular velocities in the dynamic equations of rotational motion. We suppose that the aerodynamic coefficients and their derivatives are constant.
We then have the following simplifications, where $C_n{}^{\alpha}$ is the derivative of the aerodynamic coefficient of normal force:

$$J_{yz}=0;\ J=J_y=J_z;\ C_z^{\beta}=-C_n^{\alpha};\ |m_y^{\beta}|=|m_z^{\alpha}|;$$

$$m_y=-|m_z^{\alpha}|\cdot\beta;\ m_z=-|m_z^{\alpha}|\cdot\alpha;$$

$$C_x=C_{\tau},\ C_y=C_n^{\alpha}\cdot\alpha,\ C_z=-C_n^{\alpha}\cdot\beta.$$

We can replace all these simplifications in the equations. We then have:

$$\begin{aligned}\ddot{\alpha}=&-\left[\frac{S}{m}\left(C_n^{\alpha}-C_{\tau}\right)+\frac{Sl^2}{J}|m_z^{\omega}|\right]\frac{q}{V}\dot{\alpha}-\\&-\left[\frac{qSl}{J}|m_z^{\alpha}|-\frac{J-J_x}{J}\omega_x^2+\frac{q^2S^2l^2}{mJV^2}\left(C_n^{\alpha}-C_{\tau}\right)|m_z^{\omega}|\right]\alpha-\\&-\frac{2J-J_x}{J}\omega_x\dot{\beta}-\left[\frac{Sl}{J}|m_z^{\varpi}|+\frac{J-J_x}{J}\frac{S}{m}\left(C_n^{\alpha}-C_{\tau}\right)\right]\frac{q}{V}\omega_x\beta+\\&+h_{\beta}\frac{J-J_x}{J}\omega_x^2+\frac{qSl}{J}|m_z^{\alpha}|\alpha_a-\frac{qSC_x}{J}\Delta y,\end{aligned}$$

$$\begin{aligned}\ddot{\beta}=&-\left[\frac{S}{m}\left(C_n^{\alpha}-C_{\tau}\right)+\frac{Sl^2}{J}|m_z^{\omega}|\right]\frac{q}{V}\dot{\beta}-\\&-\left[\frac{qSl}{J}|m_z^{\alpha}|-\frac{J-J_x}{J}\omega_x^2+\frac{q^2S^2l^2}{mJV^2}\left(C_n^{\alpha}-C_{\tau}\right)|m_z^{\varpi}|\right]\beta-\\&-\frac{2J-J_x}{J}\omega_x\dot{\alpha}-\left[\frac{Sl}{J}|m_z^{\omega}|+\frac{J-J_x}{J}\frac{S}{m}\left(C_n^{\alpha}-C_{\tau}\right)\right]\frac{q}{V}\omega_x\alpha+\\&+h_{\alpha}\frac{J-J_x}{J}\omega_x^2+\frac{qSl}{J}|m_z^{\alpha}|\beta_a-\frac{qSC_x}{J}\Delta z.\end{aligned}\qquad(2)$$

Where: α_a, β_a are balancing corners of attack and slip, caused by a small asymmetry of the form, $h_{\alpha}=\frac{J_{xz}}{J-J_x}, h_{\beta}=\frac{J_{xy}}{J-J_x}$, are relative centrifugal moments of inertia.

3 Modelling part and analysis

With the equations determined in the first part, we can write program which that give us data on the landing. We will focus only on one stage of the landing, the last one: the landing on the planet surface. We consider several options for approaching the surface: it is a multifactorial problem. First, we consider that the deviation of the vertical axis of the landing vehicle can be 10 degrees, 20 degrees or 30 degrees. Then, for each of these deviations, we model the landing on a hard surface (such as rock soil) and on a soft surface (such as sand). We then have 6 landing options to analyze. The conditions of the landing are the Earth conditions. Finally, we consider the following data:

- Approach speed to the surface (vertical component of speed): 35 m/s.
- Horizontal component of speed (simulating wind): 10 m/s.

The modelling give us as results the values of the overload along each axis and the total value of the overload for the apparatus, and also for the container with payload that is located inside the landing vehicle. This is landing vehicle like MetNet projects [6].
Figure 1 shows the main stages of the movement: I - Atmosphere entry. Opening primary inflatable braking device; II - Entering the additional inflatable braking device; III - Braking with the additional inflatable braking device; IV - Landing vehicle touch the ground.

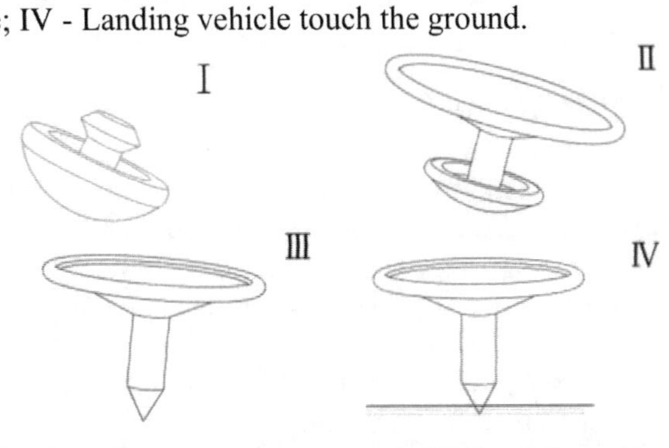

Figure 1. Descent scheme of landing vehicle movement

With this data, we can draw curves of the overload. The final goal is to determine the most optimal and the most dangerous approach options.

3.1. Modelling on soft soil (sand)

In this part, we will study the influence on the deviation of the vertical axis of the landing vehicle when it lands on soft soil, such as sand (Figure 1, Stage IV). Let's compare the overload in each of these situations. At first, we focus on the apparatus itself.

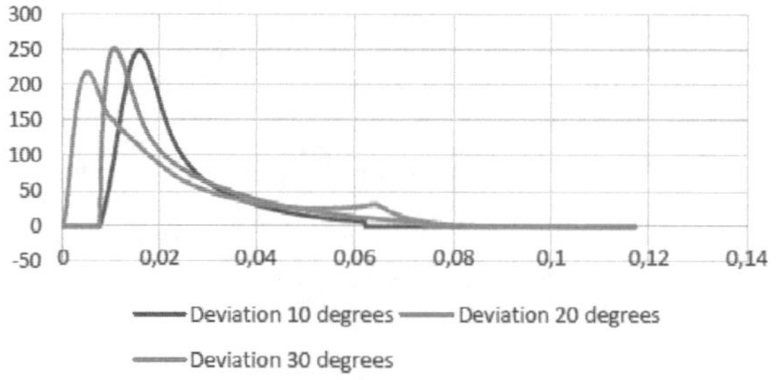

Figure 2. Overload of the apparatus when landing on soft soil

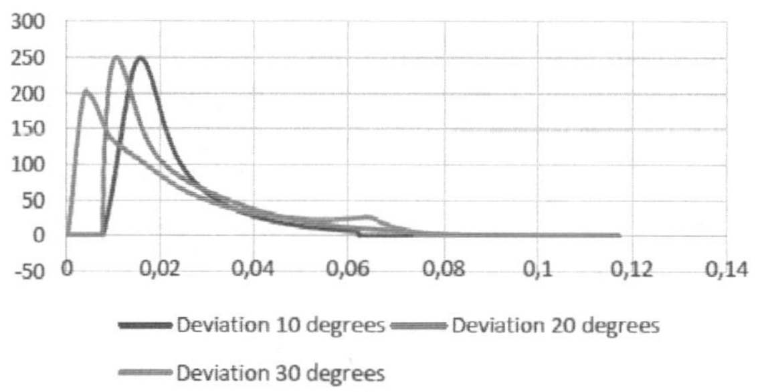

Figure 3. Overload of the apparatus on the X axis when landing on soft soil

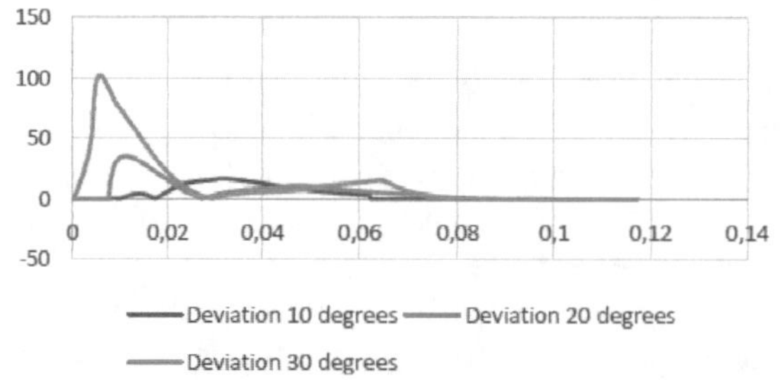

Figure 4. Overload of the apparatus on the Y axis when landing on soft soil

What is important to look at when comparing these graphics is the maximum of the overload and the duration in which the overload is non-zero.

On the Figure 2, we can see that the amplitude of the overload with a deviation of 30 degrees is lower than for the two other curves, but it presents another peak after. This last occurrence is not a good thing. The peak for a deviation of 20 degrees has the same amplitude than the one of 10 degrees, but lasts longer. The situation with an approach of 10 degrees seems to be the best.

On the Figure 3, the 10 degrees deviation and the 20 degrees deviation curves have the same amplitude, but the 20 degrees curves lasts longer. The 30 degrees curve has a lower amplitude, but still has the second peak, that is not good for the apparatus.

On the Figure 4, the amplitude of the 30 degrees curve is much higher than the one of two other curves. The 20 degrees deviation curve has a higher amplitude and lasts longer than the 10 degrees deviation one.

So, when landing on a soft soil and considering the entire apparatus, the optimal approach option is a 10 degrees deviation of the vertical axis of the landing vehicle, and the most dangerous one is the 30 degrees deviation of the vertical axis of the landing vehicle. However, the results of a 20 degrees deviation are quite close to the one obtained with a 10 degrees deviation.

Let's see if the results are the same in the container with payload that is located inside the landing vehicle. For the container, we have the following graphics.

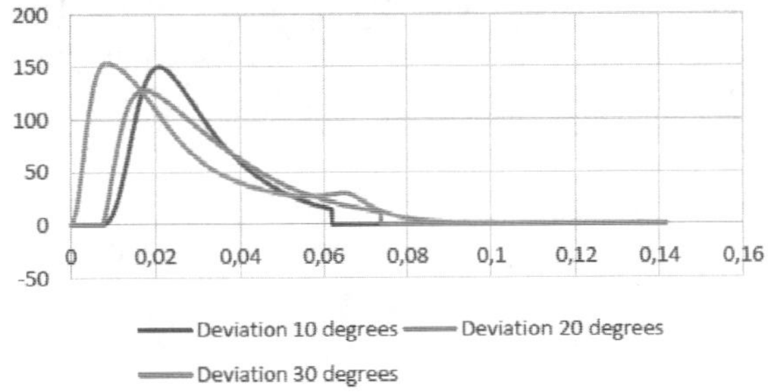

Figure 5. Overload of the container when landing on soft soil

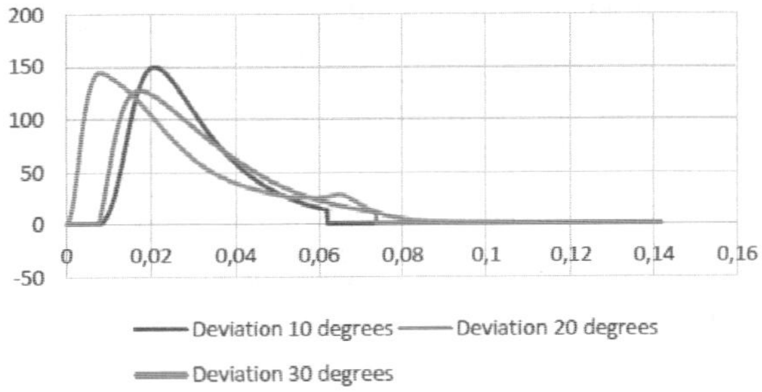

Figure 6. Overload of the container on the X axis when landing on soft soil

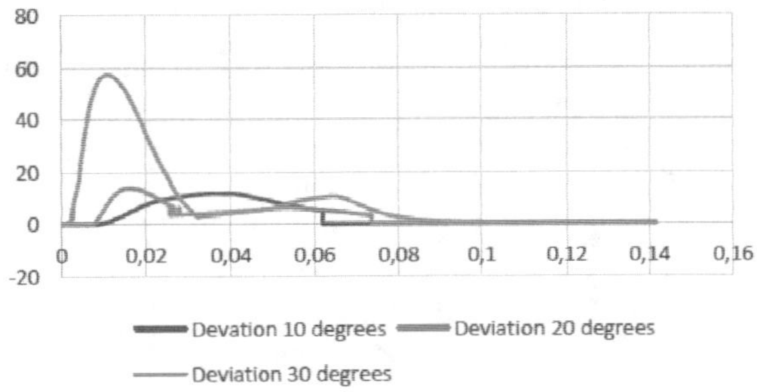

Figure 7. Overload of the container on the Y axis when landing on soft soil

On the Figure 5, we can see that the curve for the 30 degrees deviation is almost the highest, is the longest and presents a second peak. The one with the lowest amplitude is the curve for 20 degrees. But this curve lasts longer than the 10 degrees one, which has the highest amplitude.

On the Figure 6, we can make the same remarks than on the Figure 5.

On the Figure 7, it is shown that the 30 degrees deviation curve has, by far, the highest peak, and presents two of them. The 20 degrees deviation curve has the same amplitude than the 10 degrees one, but last longer.

So, in the container, when landing on a soft soil, the most optimal approach option is a 10 degrees deviation of the vertical axis of the landing vehicle, and the most dangerous one is the 30 degrees deviation of the vertical axis of the landing vehicle.
We can then conclude that, when the vehicle lands on a soft soil, such as sand, it is more optimal for it, when focusing on the overload of the apparatus and of the container with payload located inside it, to have a 10 degrees deviation of its vertical axis. It is also more dangerous to land with a 30 degrees deviation of its vertical axis.

3.2. Modelling on hard soil (rocks)
Now we will study the results of the simulations of landing on a hard soil, such as rocks, and compare the overload in the different deviation situations. Let's focus first on the apparatus.

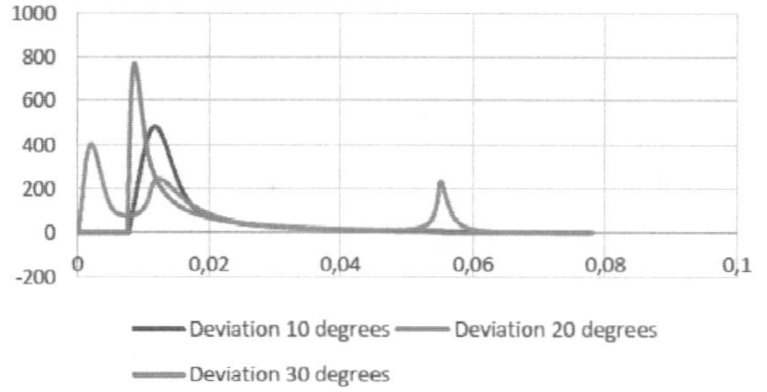

Figure 8. Overload of the apparatus when landing on hard soil

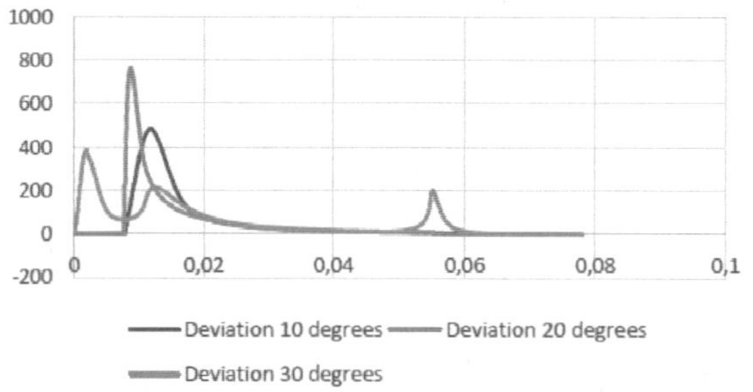

Figure 9. Overload of the apparatus on the X axis when landing on hard soil

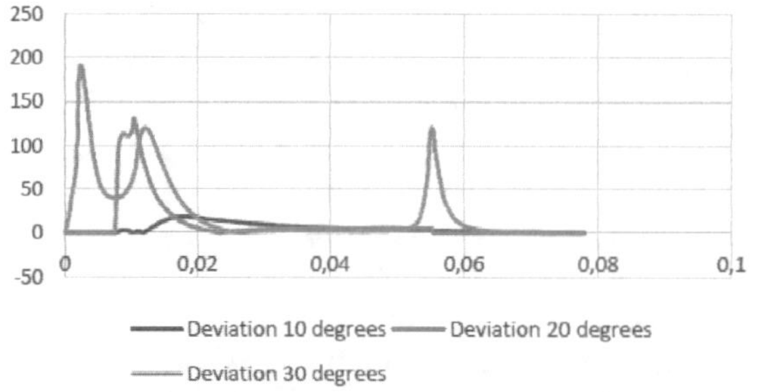

Figure 10. Overload of the apparatus on the Y axis when landing on hard soil

On the Figure 8, we can see that the 30 degrees deviation curve present three peaks, which amplitudes are not the highest, but the duration of the non-zero overload is very long. The 10 degrees deviation curve and the 20 degrees deviation have the same duration time and present only one peak, but the peak of the 20 degrees deviation one is much higher.
On the Figure 9 and 10, we can make the same remarks, as for the Figure 8.

So, when landing on a hard soil and considering the entire apparatus, the optimal approach option is a 10 degrees deviation of the vertical axis of the landing vehicle, and the most dangerous one, by far, is the 30 degrees deviation of the vertical axis of the landing vehicle.
Let's see if the results are the same in the container with payload that is located inside the landing vehicle. For the container, we have the following graphics.

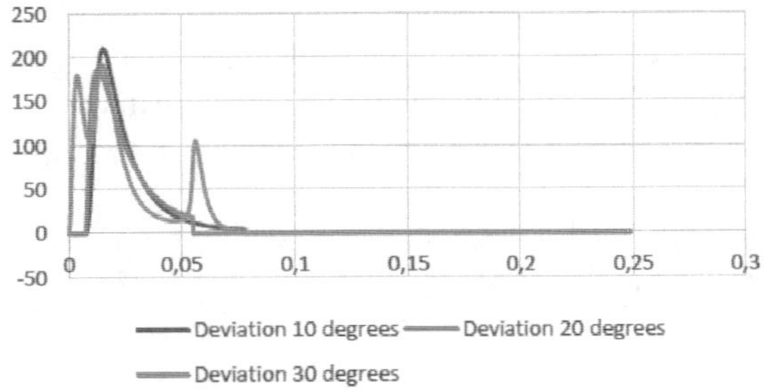

Figure 11. Overload of the container when landing on hard soil

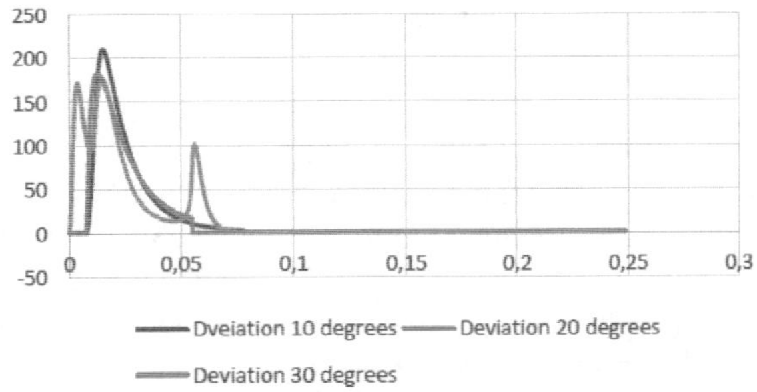

Figure 12. Overload of the container on the X axis when landing on hard soil

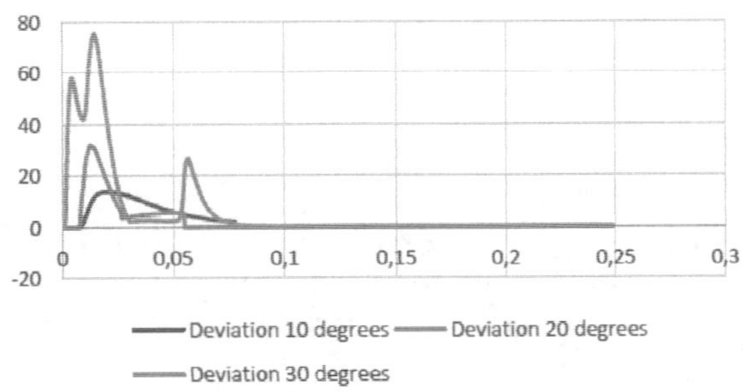

Figure 13. Overload of the container on the Y axis when landing on hard soil

On the Figure 11, we can see that the 30 degrees deviation curve presents three peaks. The maximum amplitude of each of the three curves are almost the same, the 10 degrees deviation one a little bit higher. The 10 degrees and 20 degrees curves last the same. On the Figure 12, we can make the same remarks.
On the Figure 13, we can see that the amplitude of the peaks of the 30 degrees deviation curve are much higher than the ones of the other curves. These curves have the same duration, but the peak on the 20 degrees deviation curve is higher.
So, in the container, when landing on a hard soil, the most optimal approach option is a 10 degrees deviation of the vertical axis of the landing vehicle, and the most dangerous one, by far is the 30 degrees deviation of the vertical axis of the landing vehicle.

We can then conclude that, when the vehicle lands on a hard soil, such as rocks, it is more optimal for it, when focusing on the overload of the apparatus and the container, to have a 10 degrees deviation of its vertical axis. It is also more dangerous, and by far, to land with a 30 degrees deviation of its vertical axis.

4. Conclusion

The paper is analyzes the movement of the landing vehicle at the final stage of its movement. For this, a system of equations was compiled describing the motion of the landing vehicle. For the composed system of equations, a special program was created that allows one to calculate the motion of such a landing vehicle.
At the final stage of the movement, at the stage of contact with the surface, one of the most important quantities arising is the magnitude of the resulting overloads. It is for this parameter, overloads, taking into account the developed algorithm, the movement was analyzed at the final stage of movement - contact with the surface.
After analyzing the data given by the algorithm, we can say as a conclusion that, during the landing of an apparatus, the deviation of its vertical axis has impact on the landing: the more important the deviation is, the more dangerous the landing become. So the optimal approach options are when the deviation of the vertical axis of the landing vehicle is the lowest, and the most dangerous approach options are when it is the highest.

References

[1] V.V. Koryanov. Method of calculating the descent of the spacecraft in the atmosphere using technology adaptation landing in different environmental conditions, IAC-14-C2.3.4, 65th Astronautical Congress, Toronto, Canada, 2014, 29 September – 03 October.

[2] V.V. Koryanov, V.P. Kazakovtsev. Dynamics of angular motion of landing vehicle in martian atmosphere with allowance for small asymmetries. *International Journal of Mechanical Engineering and Robotics Research*, Volume 7, Issue 4, 1 July 2018, Pages 385-391.

[3] Michele Iacovazzo, Valerio Carandente, Raffaele Savino, Gennaro Zuppardi. Longitudinal stability analysis of a suborbital re-entry demonstrator for a deployable capsule. *Acta Astronautica. Volume 106*, January–February 2015, pp. 101–110.

[4] Carandente, Valerio, et al. A Study on Earth Re-entry Capsules with Deployable Aerobrakes for Recoverable Microgravity Experiments. *Microgravity Science and Technology* 27.3. 2015. 181-191.

[5] Zuppardi, G., R. Savino, and G. Mongelluzzo. Aero-thermo-dynamic analysis of a low ballistic coefficient deployable capsule in Earth re-entry. *Acta Astronautica* 127 (2016): 593-602.

[6] Jyri Heilimo, Ari-Matti Harri, Sergey Aleksashkin, Vsevolod Koryanov, Ignacio Arruego, Walter Schmidt, Harri Haukka, Valery Finchenko, Maxim Martynov, Boris Ostresko, Andrey Ponomarenko, Viktor Kazakovtsev, Susanna Martin, and Tero Siili. RITD - Adapting Mars Entry, Descent and Landing System for Earth. Geophysical Research Abstracts Vol. 16, EGU2014-5506-1, 2014, EGU General Assembly 2014

[7] The Mars MetNet Mission. Concept Drawing. http://fmispace.fmi.fi/old-metnet/index.php?id-100, (accessed: 05.02.2019).

[8] Pichkhadze K.M., Finchenko V.S., Aleksashkin S.N., Ostreshko B.A. Transformable vehicles descending in the atmospheres of the planets, Herald Federal State Unitary Enterprise Scientific and Production Association named after Lavochkin, 2. 2015, 4-13.

[9] V.S. Zarubin, V.N. Zimin, G.N. Kuvyrkin. Temperature Distribution in the Spherical Shell of a Gauge-Alignment Spacecraft. *Journal of Applied Mechanics and Technical Physics* (2017) 58. pp. 1083-1090. https://doi.org/10.1134/S0021894417060141.

[10] V.V. Leonov. The design features of inflatable large-scale mirror concentrators for space high-temperature solar power plant. Proceedings of the international astronautical congress, IAC 67, Making Space Accessible and Affordable to All Countries. 2016.

[11] N.A. Nerovny, I.E. Lapina, A.S. Grigorjev. Light radiation pressure upon an optically orthotropic surface. *Journal of Quantitative Spectroscopy and Radiative Transfer*. Volume 202, November 2017, pp. 64-73. https://doi.org/10.1016/j.jqsrt. 2017.07.016.

[12] Nerovny, N.A., Zimin, V.N. Light radiation pressure upon a wrinkled membrane - Parametrization of an optically orthotropic model. Journal of Physics: Conference Series Volume 991, Issue 1, 13 April 2018, Article number 01206 25th International Conference on Topical Problems of Continuum Mechanics, TPCM 2017; Tsakhkadzor; Armenia; 2 October 2017.

[13] Canuto E. et al. Planetary landing: Modelling and control of the propulsive descent //*Proceedings of the 31st Chinese Control Conference*. – IEEE, 2012. – C. 7309-7316.

The dynamical analysis of the modified rossler system

Aleksandra Tutueva, Denis Butusov, Artem Okhota, Dmitrii Pesterev and Ekaterina Rodionova

Saint Petersburg Electrotechnical University "LETI", 197376 Saint Petersburg, Russia

avtutueva@etu.ru

Abstract. The synthesis of novel chaotic systems is a modern branch of nonlinear dynamics since deterministic chaos properties can be successfully applied in various engineering and scientific problems. In this paper we investigate changes in the dynamics of the modified Rossler system after applying coordinates transformation to the original model. We perform the bifurcation analysis of the obtained model and experimentally show that its behavior differs with the behavior of the prototype. We study the finite-difference schemes obtained for considered chaotic systems and find their similarity in simulation with different integration steps. We show that discretization effects are the source of the small differences between the two models. The obtained results can be used in theoretical nonlinear dynamics, nonlinear systems simulation, development of communication and control systems.

1. Introduction

Chaotic dynamics can be observed in electrical circuits [1–4], liquid or gas convection [5], motion trajectories of celestial bodies [6], a human heartbeat [7] and many others. Identification and improvement of mathematical models of such processes is an important task. On the other hand, sensitivity to initial values and topological mixing allow to efficiently apply chaotic systems in control, security, and communication systems [8–10]. Thus, the synthesis of chaotic systems with the required properties is an actual topic of nonlinear science. In [11] methods for obtaining symmetric discrete maps with an adaptive phase space were proposed. Furthermore, methods of controlling the number of equilibria and a Lyapunov exponent value were considered in [12–14]. Recently, Nepomuceno et al. proposed an easily implementable digital model with chaotic behavior [15]. Such studies usually thoroughly describe the technique of synthesizing new mathematical objects from a well-studied system leaving in the background the questions of the developed models adequacy and the influence of a discrete operator. Nevertheless, it was repeatedly shown that even the order of performing mathematical operations can significantly affect its dynamics while a chaotic system is simulated [16].

In our study we experimentally compare the behavior of the well-known chaotic Rossler system with its modified version obtained using an affine coordinates transformation [12]. We expect, that the applied synthesis method will not introduce significant changes in the system dynamics excluding the discretization effects impact.

The paper is organized as follows. In Section II we describe a mathematical model of the modified Rossler system. Bifurcation analysis and the Largest Lyapunov exponent calculation are carried out in Section III. Section IV shows experimental study of numerical integration methods applied to considered systems. Finally, some conclusions are given in Section VI.

2. The double-Rossler chaotic system

In [12] X. Wang and G. Chen proposed the technique for constructing system with any number of equilibria. The proposed method is based on a coordinate transformation that allows to obtain the phase space symmetry relative to a chosen axis. The number of equilibrium points is controlled by the rotation angle. Let us consider well-known Rossler chaotic system [17]

IOP Conf. Series: Materials Science and Engineering **630** (2019) 012006 doi:10.1088/1757-899X/630/1/012006

$$\begin{cases} \dot{x} = -y - z; \\ \dot{y} = x + ay; \\ \dot{z} = b + z(x - c). \end{cases} \qquad (1)$$

where *a*, *b*, *c* are nonlinearity parameters. The Rossler system has one saddle-focus equilibrium point with 2D unstable manifold and another one saddle-focus equilibrium point with 1D unstable manifold (Fig. 1).

Rewrite system (1) in terms of *u*, *v* and *w* as follows:

$$\begin{cases} \dot{u} = -u - w; \\ \dot{v} = u + av; \\ \dot{z} = b + w(u - c). \end{cases} \qquad (2)$$

Consider the following simple coordinate transformation:

$$\begin{cases} u = x^2 - y^2; \\ v = 2xy; \\ w = z. \end{cases} \qquad (3)$$

Transformation (3) can add a *z*-axis rotation symmetry $R_z(\pi)$ to the original system (1) because for each (*u*, *v*, *w*) there are two points (±*x*, ±*y*, ±*z*) corresponding to (*u*, *v*, *w*).

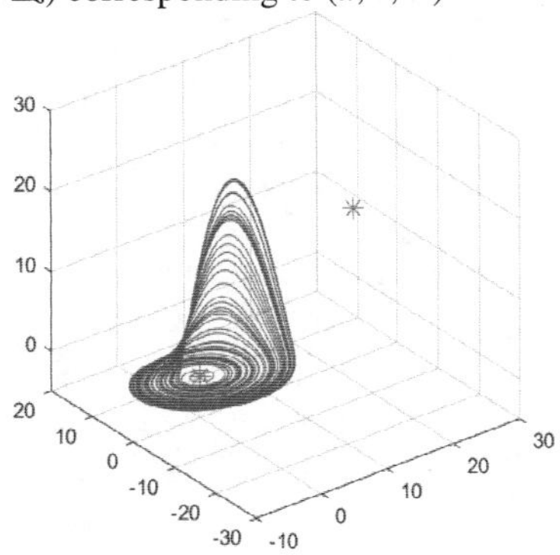

Figure 1. The phase portrait and equilibrium points of original Rossler system

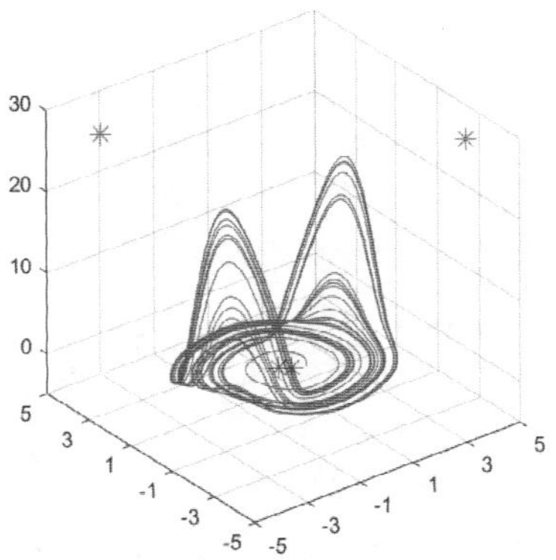

Figure 2. The phase portrait and equilibrium points of the double-Rossler system

After the transformation (3), the system becomes

$$\begin{cases} \dot{x} = \dfrac{-x^2y - zx - y^3 + 2axy^2}{2(x^2 + y^2)}; \\ \dot{y} = \dfrac{xy^2 + zy + x^3 + 2ax^2y}{2(x^2 + y^2)}; \\ \dot{z} = b + z(x^2 - y^2 - c). \end{cases} \qquad (4)$$

IOP Conf. Series: Materials Science and Engineering **630** (2019) 012006 doi:10.1088/1757-899X/630/1/012006

We call the resulting system as the *double-Rossler system* since it possesses two pairs of equilibrium points symmetrical by the z-axis (Fig. 2). As in the original model, one equilibrium point is located in the middle of the attractor and is a saddle-focus with an unstable 2D manifold. Second equilibria is outside of the region of the attractor and is saddle-focus equilibrium point with 1D unstable manifold.

Let us consider the results of the dynamical analysis of the obtained model in comparison with the original Rossler system.

3. Dynamical analysis

3.1. Bifurcation analysis

Bifurcation analysis of the chaotic Rossler system usually is carried out with fixed values $a = 0.2$ and $b = 0.2$ while c is varied. Within $2.6 \leq c \leq 4.2$, the system have a stable limit cycle. In this interval there is a period doubling. Then, after $c = 4.2$ chaotic behavior can be observed.

In this study to plot the bifurcation diagrams we simulate the considered systems on the time interval $t = 700$ sec using the semi-explicit numerical integration *SED*-method [18] with integration step $h = 0.01$. We use variable z to construct the Poincaré sections. The bifurcation diagrams for both considered systems are shown in Fig. 3.

One can see that the behavior of the double-Rossler model is similar to the original system while increasing parameter c. The most significant difference can be noticed in the so-called windows of bifurcation diagrams, corresponding to the intervals [5.3; 5.6], [6.7; 6.9] and [7.8; 8.2]. This mismatch is also confirmed by the values of the Largest Lyapunov exponent (LLE). In the original model LLE tends to a negative value near $c = 8$ which corresponds to the harmonic mode of oscillation. In the double-Rossler system in this interval a chaotic mode is observed. It can be assumed that a variation of models behavior is related to the appearance of the effects from the discretization method. Let us consider methods for obtaining finite difference schemes of the investigated chaotic models and compare them.

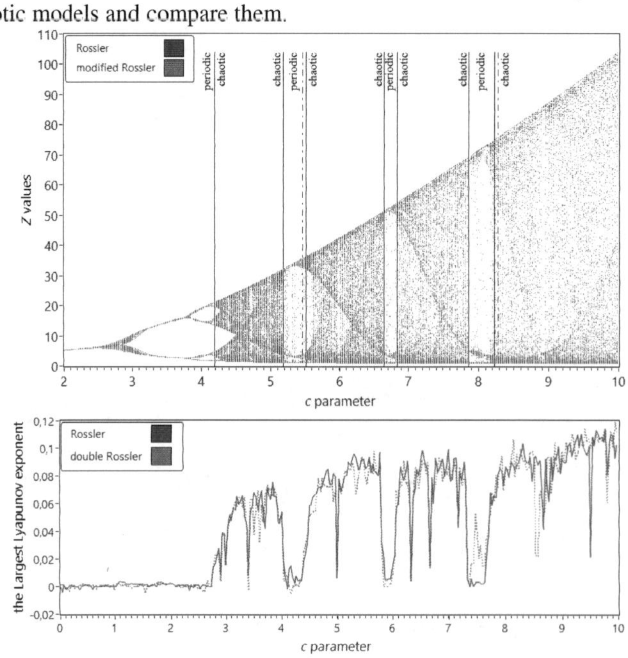

Figure 3. Bifurcation diagrams and LLE for Rossler and double-Rossler systems

3.2. Analysis of finite difference schemes

The study of discretization methods is performed using so-called step bifurcation diagrams. If we choose the integration step as the nonlinearity parameter, the diagram is called a step-diagram or h-diagram. We consider the common techniques of integration ordinary differential equations including explicit and implicit midpoint methods and the semi-implicit *SED*-method. The experimental results are shown in Fig. 4–6.

The simulation of investigated systems using the explicit midpoint method shows the differences between two considered models. For the double-Rossler system chaotic modes appear in the interval [0.1; 0.15] in contrast with the original system. In diagrams obtained for the implicit midpoint method and the *SED*-method the windows of bifurcation diagrams are shifted towards an integration step increasing. Thus, one can conclude that the possible source of the different behavior of the Rossler models is the discretization method. Furthermore, the data type can significantly affect the discrete model due rounding the result of a calculation. Since the equations after the coordinate transformation are more complex than the original Rossler system, then it is worth using high precision floating-point data types and high-order integration methods to achieve the required model adequacy.

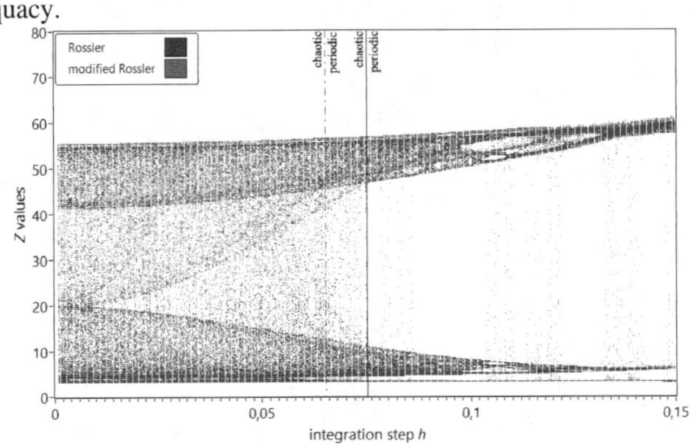

Figure 4. *h*-diagrams of explicit midpoint method for Rossler and double-Rossler systems

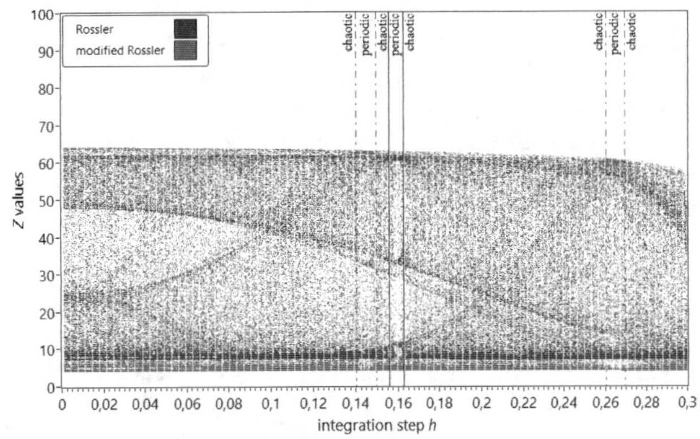

Figure 5. *h*-diagrams of implicit midpoint method for Rossler and double-Rossler systems

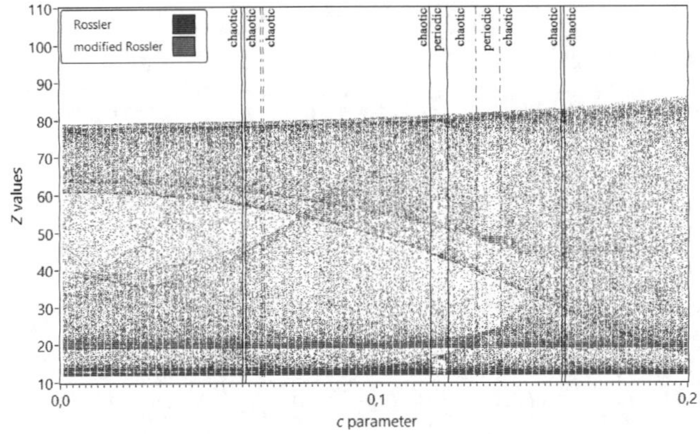

Figure 6. *h*-diagrams of semi-implicit *CD*-method for Rossler and double-Rossler systems

4. Conclusions

In this paper we considered the modified Rossler system obtained using coordinates transformation of the original model. We show that using the same integration method for discretization of differential equations yields similar, but still different results. The additional investigation was carried out to reveal the reasons of such behavior. Using the step-diagrams, we explicitly show that the main source of misbehavior is the applied discrete operators. Thus, one should be extremely careful when constructing novel chaotic systems using coordinate transformation approach.

The topic of our further studies will be the investigation of various models of chaotic systems obtained through coordinate transformations and the detailed study of numerical method impact. The most interesting case to investigate are conservative chaotic systems that are significantly different from dissipative counterparts. We will study the various oscillations modes arising in such systems and the property of preserving the phase volume during long-term simulations.

References

[1] Z. Chen, X. Zhang, Q. Bi, Nonlinear Anal. Real World Appl., **9.3**, 1158-1168 (2008)
[2] C. Zheng, H. Iu, T. Fernando, D. Yu, H. Guo, J. K. Eshraghian, Chaos, **28.6**, 063115 (2018)
[3] F. Yuan, G. Wang, P. Jin, X. Wang, G. Ma, IJBC, 26.08, 1650130 (2016)
[4] L. O. Chua, M. Itoh, L. Kocarev, K. Eckert, J. Circuit Syst. Comp., **3.01**, 93-108 (1993)
[5] E. N. Lorenz, JAS, **20.2**, 130–141 (1963)
[6] V. S. Andreev, S. V. Goryainov, A. V. Krasilnikov, *Proceedings of 2016 XIX IEEE International Conference on Soft Computing and Measurements (SCM)*, 196–198 (2016)
[7] D. Creanga, C. Nadejde, P. Gasner, Rom. J. Phys **56.1–2**, 177–185 (2011)
[8] A. V. Tutueva, D. N. Butusov, D. O. Pesterev, D. A. Belkin, N. G. Ryzhov, *Proceedings of 2017 International Conference «Quality Management, Transport and Information Security, Information Technologies» (IT&QM&IS)*, 284–287 (2017)
[9] T. L. Vincent, J Nonlinear Dynamics and Systems Theory **1(2)**, 205–218 (2001)
[10] T. Karimov, D. Butusov, V. Andreev, A. Karimov, A. Tutueva, Electronics, **7.7**, 123 (2018)
[11] D. N. Butusov, A. I. Karimov, N. S. Pyko, S. A. Pyko, M. I. Bogachev, Physica A, **509**, 955–970 (2018)
[12] X. Wang, G. Chen, Nonlinear Dyn., **71.3**, 429–436 (2013)
[13] X. Zhang, Int. J. Bifurc. Chaos, **27.08**, 1750118 (2017)
[14] C. Wang, C. Fan, Q. Ding, Int. J. Bifurc. Chaos, **28.07**, 1850084 (2018)
[15] E. G. Nepomuceno, A. M. Lima, J. Arias-Garcia, M. Perc, R. Repnik, Chaos Solitons Fractals, **120**, 62–66 (2019)
[16] E. G. Nepomuceno, H. M. R. Junior, S. A. M. Martins, M. Perc, M. Slavinec, Appl Math Comput **336**, 67-75 (2018)
[17] O. E. Rössler, Phys. Lett. A, **57(5)**, 397–398 (1976)
[18] D. N. Butusov, V. Y. Ostrovskii, A. I. Karimov, V. S. Andreev. IJERTCS. **10(2)** (2019)

IOP Conf. Series: Materials Science and Engineering **630** (2019) 012007 doi:10.1088/1757-899X/630/1/012007

Dialectical Understanding of Information in the Context of the Artificial Intelligence Problems

Ibragim E Suleimenov[1], Oleg A Gabrielyan[2], Akhat S Bakirov[1] and Yelizaveta S Vitulyova[1]

[1] Almaty University of Power Engineering and Telecommunications, Almaty, Republic of Kazakhstan
[2] V.I. Vernadsky Crimean Federal University, Simferopol, Russian Federation

axatmr@mail.ru

Abstract. It is shown that the implementation of breakthrough progress in artificial intelligence cannot be accomplished without a radical modernization of the methodological framework for research in this area. This upgrade can be performed only with the use of conceptual and methodological apparatus of dialectics. Its consistent application to objects, in either case connected with ideas about the intellect, results in the formulation of the dialectic symmetry principle. The first practical steps for the implementation of the proposed concept in practice are associated with the creation of a new education paradigm, which operates, including ideas that "education" being understood as a public institution, is not reducible to individual learning. Education in modern conditions can and should be viewed primarily as a tool to influence public consciousness, or more precisely, its subsystems.

1. Introduction

The problem of artificial intelligence (AI) is becoming one of the main areas of research related to the emergence of the digital age [1]. The most controversial issue that arises when discussing the essence of AI, obviously, is the issue of the nature of intelligence as it is. Currently, this issue is not abstract and theoretical but of immediate practical significance, including in connection with the development of tools to adequately manage the information flow in society.

They have repeatedly tried to give an answer to it from various positions [2],[3], but all these have very definite disadvantage associated with the fact that the concept of "intellect" is inseparably linked with the concept of "information", which until quite recently has had no consistent interpretation in the literature.

Indeed, the "intellect" certainly is a kind of essence of a purely informational nature. Thus, the human mind is a result of the exchange of signals between neurons in the brain, each of which performs relatively simple functions. We say that consciousness is a system property, but in itself this statement only states a fact that is obvious to date - it is not reducible to the properties of individual neurons, it is a new quality that appears in a complex system due to the presence of an extensive system of connections between its elements.

We do not put a sign of identity between "intellect" and "consciousness". As a working model for our reasoning, we use the approach of D. Dennett. It represents consciousness as an integrated system of psychological states of a person, mutually related with each other. New information is correlated with the previously obtained. Intellect does not exhaust consciousness completely [4].

In this paper, an attempt is made to show that human consciousness is the result of self-organization processes, only indirectly related to a material carrier; it is a self-organization of certain entities that have a purely informational nature. We emphasize that at the present time it is already possible to say with certainty that in many complex systems of different nature, non-trivial informational entities that are only indirectly connected with material information carriers arise in many complex systems of different nature. Such informational entities have their own behavior, and it is often difficult to detect the manifestations of their properties by studying the behavior of the elements of the system separately. An example of a non-trivial informational entity of this kind is bureaucracy - the result of the processes of informational self-organization in administrative systems. As shown in [5], bureaucracy is a kind of remote equivalent of a biological organism, which has a pronounced equivalent of the self-preservation instinct and its own "goal setting", the character of which, as a rule, has nothing to do with the tasks for which the administrative apparatus was originally created.

This conclusion is confirmed by simple mathematical models, in particular, the voting model in the academic (or other) council [6],[7]. In this model, each member of the voting board is assigned to a separate neuron, and the voting procedure is considered as a change in its output state ("In favour", "Against", "Abstained") depending on the array of information received at its input (for example, the content of the report thesis). In the cited work it is proved that, provided that the density of connections between the elements of the system is sufficiently high, the de facto decision is made not by the set of council members, but by the neural network formed by them. (The nature of relations between council members is determined by their mutual influence on each other; for example, a council member can vote "Against" if a student of his opponent defends himself even when the dissertation is quite decent).

2. Dialectical understanding of information

Studying the processes of self-organization of entities of informational nature is a nontrivial task, since they are unobservable. Any conceivable experiment, one way or another, can provide information only about changes in the state of material bodies - information carriers, it is possible, therefore, to judge the processes of self-organization of information entities only from indirect data. (In particular, this applies to situations where information is distributed, as is the case in neural networks).

When studying the behavior of entities of informational nature, it is methodology that comes to the fore. Indeed, you need to have a theoretical tool that will allow you to adequately interpret a large array of experimental data and bring them into a form that allows you to judge what is happening in the communication shell of a complex system. The basis for the development of such tools can obviously be only a methodology that is obviously insufficiently developed for information theory, because until very recently there was no adequate interpretation of the very concept of "information" as such.

As shown in [8], this interpretation can only be given by means of philosophy, since the concept of information is among the fundamental. More precisely, this concept can and should be viewed precisely as a category of dialectics. In particular, this means that the study of problems associated with the concept of "intellect", in particular, the most fundamental problems of artificial intelligence, cannot be conducted otherwise than with the involvement of an apparatus of objective dialectics.

Hence there is a need to create our own methodological basis for the theory of artificial intelligence, which is fundamentally different from the one that is currently in use. In fact, the methodology of work carried out in this area does not differ from the methodology used in the development of classical computing devices. Obviously, sooner or later it will be necessary to go beyond it, since AI systems must necessarily be something qualitatively different than existing digital devices.

The basis of the proposed methodology can be briefly defined through the "dialectic symmetry principle". Consider the grounds for the promotion of this principle.

As noted in [8], information is one of the fundamental concepts widely used in both the natural and human sciences. Therefore, talking about the "definition" of information in the sense of school definition seems to be unjustified. To define in such a context is to reveal the content of one term using others. Such reductionism does not work for fundamental concepts.

Consequently, there must be certain basic categories that are not disclosed through any other, i.e. to give them a definition in the classical sense of the word is really impossible because of their fundamental nature. Dialectics defines such categories through opposition.

There are a number of paired dialectic categories: content and form, quantity and quality, etc. They do not have the usual, "school" definition precisely because of their fundamental nature.

These concepts include, in particular, the category of matter and information [8]. It is curious to note that the philosophical interpretation of the category "matter" experiences the same difficulties as the definition of the "information" category, despite its historical background.

In materialistic philosophy, it is defined as follows. Matter is a philosophical category for the designation of objective reality, which is given to man in his sensations, is displayed by our sensations, existing independently of them. Or "from the point of view of dialectics, matter is objective reality - the reason, the basis, the content and the carrier of the whole diversity of the world" [9].

Until recently, in the materialist dialectics, the category of matter has occupied a special place, i.e., not having a pair. Realizing that the basic categories cannot be defined in any other way except through opposition, the experts on this issue left the category of "matter" (in this sense) aside.

Difficulties are removed automatically, if it is recognized that "matter" and "information" are paired dialectic categories. They, like others, should be defined only through opposition. There is a point of view according to which information is contrasted with energy [10]. However, it cannot be considered consistent, since motion is a form of existence of matter.

The nature of objective reality is dual, which in one sense or another has been recognized and is still recognized by all schools of thought. Any objectively existing object contains (we state it this way, for lack of a better term) both material and informational components. For example, by studying specific water samples, you can get information about water as such. Simplifying, in a separate water molecule there is information about all such molecules.

The proposed approach, which can be called the principle of the dualism of matter and information, automatically puts everything in its place. However, using the opposition formulated above, at first glance, the constructiveness of the definition of information disappears; it becomes too general to be used for practical purposes.

The constructiveness of the definition of information given through opposition is expressed in that on this basis it is possible to formulate the dialectic symmetry principle. (As it is known, many important scientific results were obtained precisely on the basis of considerations of symmetry, for example, the Onsager principle).

Similarly, the principles of symmetry can also be applied to the opposition of matter and information, considered as paired dialectic categories. First of all, it follows from this principle that just as there can exist an infinite variety of forms of matter, just as there can exist an infinite variety of forms of information. Matter can be inert, and can also be alive, therefore, symmetrical concepts must exist in relation to information; This creates additional opportunities for the study of nontrivial informational entities arising in complex systems by philosophical methods.

Further, it is precisely the principle of dialectic symmetry, associated with the duality of matter and information, that makes it possible to adequately interpret the relationship between what is called "energy" and what is called 'information". As you know, there are numerous speculations about the question of whether information can have "energy" and the like. Refer to table 1.

TABLE 1. TO ILLUSTRATE THE DIALECTIC SYMMETRY PRINCIPLE.

Substance	⟺	Energy
Information	⟺	?

"Factor X", indicated by a question mark in this table, has no name yet and, apparently, has not even been discussed in the scientific literature (in any case, it has not been widely discussed). However, judgments related to this factor already exist in the mass consciousness. So, it is customary

IOP Conf. Series: Materials Science and Engineering **630** (2019) 012007 doi:10.1088/1757-899X/630/1/012007

to talk about a person's positive or negative "energy", mental energy, there are also ideas that information can somehow be converted into energy and/or be considered as its form.

Analyzing the entire spectrum of such judgments is obviously meaningless, since nothing stands behind them, except for certain intuitive considerations that remain unprovable.

You can do it easier - remember the genesis of the concept of "energy" in its modern interpretation and try to act - or outline a plan of action – in a similar way.

If we turn to the history of science, we find that the concept of "kinetic energy" was first used not long ago, in 1829 (the work of G. Coriolis). The concept of "potential energy" is even later - 1853 (U. Renkin); by historical standards it is the most recent.

Initially, both of these concepts appeared as the result of a mathematical study of equations describing the motion of a material point. What later became the formula for kinetic energy — the product of mass and the square of velocity — appears as the first integral in solving a second-order differential equation that does not explicitly include the time variable. The situation is similar with potential energy - the corresponding function was introduced just to somehow name the result of the implicit integration of a vector function describing the force field.

Echoes of mid-XIX century discussions, whose participants - from different positions - tried to give an interpretation of elegant mathematical constructions leading to formulas, later interpreted as an expression of conservation laws, can also be traced in philosophical literature - it is enough to recall Anti-Dühring of F. Engels. It is appropriate to recall these discussions here for a very definite reason - the concept of "energy" in its modern form was born from mathematical calculations and only then received a physical and philosophical interpretation. It penetrated into mass consciousness much later - somewhere at the turn of XIX and XX centuries.

Consequently, there is every reason to raise the question of building a theory, which will include information that has the ability to act - parallels with the early ideas about "living force" and "kinetic energy" can be seen quite clearly.

Apparently, such considerations in the foreseeable future will form the basis for a quantitative description of the behavior of non-trivial information structures generated by complex systems, which, among other things, have an increasingly pronounced effect on human society.

The principle of dialectic symmetry can be applied not only to clarify the issue of "information energy". From it, in particular, directly implies the existence of what above was called non-trivial information entities. Refer to Table 2.

TABLE 2: THE HIERARCHY OF LEVELS OF ORGANIZATION OF MATTER AND INFORMATION.

Levels of organization of matter	**Levels of organization of information**
Inorganic level Sub Levels: - microworld: submicro-elementary - microelementary - nuclear - atomic - molecular levels - macroworld: macroscopic bodies - megaworld: planets - star-planetary complexes - galaxies - metagalaxies	Inalienable information, information alienated as a result of physico-chemical interactions
Biological level Sub Levels: - level of biological macromolecules - cell level - microorganism level - level of organs and tissues - level of the organism as a whole - population level	Relatively isolated information with its own behavior, self-organizing information

- biocenosis level - biosphere level	
Social level Sub Levels: - level of individuals - family level - level of teams - level of social groups - level of nationalities and nations - the level of states and systems of states (civilizations) - universal (noosphere) level	Information that is aware of itself, the essence of the highest information of different nature

This table is designed to emphasize the following circumstance. Just as a matter can appear in the most diverse manifestations, so information cannot be regarded as homogeneous (in particular, there is not and cannot be a complete understanding of information relating to all its manifestations). From the principle of dialectic symmetry it follows unambiguously that just as there are hierarchical levels corresponding to the dialectical category of matter, so there are levels corresponding to the category of information.

In particular, we are talking about non-trivial informational entities arising and developing in the communication space, mentioned above. The term "public consciousness" is still widely used in the post-Soviet humanitarian literature. The interpretation of this term, available in philosophical dictionaries and reference books, is de facto completely built on denials or is a tautology, or expresses the concepts in question through other, conceived less clear. One of them sounds, for example, so the public consciousness is "a display in the spiritual activity of people of interests, representations of various social groups, classes, nations, society as a whole" [11]. As noted in [12], the application of such a definition is somewhat difficult due to the vagueness of the concept of "spiritual activity".

In M. Mamardashvili we find the following definition. Social consciousness is a reflection of social being; set of collective representations inherent in a particular era. It reflects in essence the very state of a particular society. Public consciousness is opposed to individual consciousness as something in common, which is contained in the consciousness of each person as a member of society. [13].

The key in the above and similar definitions is the fact that the social consciousness (no matter what this terms means) cannot be reduced to a set of individual consciousness, that is - something else. However, such statements do not allow in the slightest degree to reveal the nature of this entity. A step in this direction was made in the monograph [14], where the neural network model of the noosphere was proposed. Within the framework of this model, the society was considered as an analogue of a neural network, and individual neurons were assigned to individuals, and interpersonal communications - connections between neurons of the network.

This analogy suggests that the informational entity, called the public consciousness, actually has a very complex structure [15]. This follows from the fact that it is generated by an extensive communication system, a very complex architecture. Moreover, there is every reason to assert that the public consciousness is formed by a combination of different informational entities, the specifics of which are determined by the characteristics of the generating communications. So, generic and family-clan ties generate informational entities of one type, professional communications - of another one, etc.

We will show it. Let us push aside the simplified model of the functioning of the brain. According to modern concepts, human intelligence and human consciousness are formed as a result of the fact that the neurons of the cerebral cortex exchange signals with each other, causing a change in the state of the neuron. These ideas were incorporated into the theory of neural networks, which are widely used nowadays for various applied purposes [16]. Moreover, it is necessary to emphasize that the theory of neural networks was initially created as an attempt to uncover the mechanisms of brain functioning, i.e. those mechanisms that lead to the emergence of intelligence and consciousness.

However, in the theory of neural networks, a number of very significant results have been obtained that are directly related to the problem under consideration. This information is distributed. The nature of the functioning of the human brain does not exist in the form of separate “memory cells”. The information is recorded by the system as a whole [17]; the neural network often has the ability to reproduce a hologram, which is the same image fragment as the original hologram, but only with degraded quality. This circumstance is more than important, since it allows one to proceed to the consideration of the global communication environment from the point of view of the mechanism of functioning of the human brain.

Indeed, any communication between individuals is actually an exchange of signals between the neurons of the brain of each of them. If two people enter into a dialogue, it is customary to say that two individuals communicate with each other. In reality, this is nothing more than an approximation. In fact, we are talking about the exchange of signals between two (or several) fragments of a common neural network, each of which is localized within the brain of each of the interlocutors. Continuing this logic, one can come to the conclusion about the existence of a global neural network. In [14], where it was first proposed to consider society from the point of view of analogy with neural networks, such a network was identified with the noosphere as a whole.

It must be emphasized here that the transition of a neuron from one state to another is in no way connected with the nature of the signals that they exchange. This shows, for example, the simplest consideration of how a person’s brain reacts to signals coming through the retina. In other words, the fact that neural network fragments exchanging signals are localized within the brain of different individuals does not affect the fact that, in aggregate, these neurons form some kind of common network.

Since the information stored by the neural network is distributed, it should be concluded that there is a very definite amount of information that is only indirectly associated with the memory of individuals, but correlates with the global neural network as a whole. This conclusion allows us to interpret such concepts as a socio-cultural code, mentality and social consciousness from purely natural-science positions, more precisely from the positions of information theory.

It is worth emphasizing that, until recently, the processes occurring in the global communication network remained practically inaccessible neither for influence, nor even for sequential study. To illustrate, it suffices to mention the following fact. Until now, one of the main methods of studying the processes occurring in society is a sociological survey. Such means, firstly, are extremely costly, and secondly, during sociological surveys, the coverage of the respondents is relatively low. The use of social online networks for similar purposes (Facebook, etc.) makes it possible to significantly expand the initial base available for direct observation.

We emphasize that the global communication environment relates to a different level of the organization of existence, rather than individuals. It is extremely difficult for an individual person to understand and comprehend the processes occurring in the global communication environment, just as it is difficult to imagine a person entering into dialogue with a single neuron from the cortex of his own brain.

The gap between these levels of the organization of things can be overcome precisely by using artificial intelligence systems, which, on the one hand, are able to process very large amounts of information, and, on the other hand, they are able to directly connect to the global telecommunications environment.

Thus, the interpretation of concepts, one way or another connected with ideas about sociocultural codes, can indeed be given on the basis of analogies with neural networks. It also creates prerequisites for the quantitative description of identities as certain information objects recorded in the global communication network - the noosphere.

Any of the scientific and technical paradigms can be considered as one of the entities that remain and develop in the public consciousness. According to T. Kuhn [17], the scientific paradigm is a set of scientific achievements recognized by the scientific community in a given period of time and serving as the basis and model for new scientific research. Essentially, Thomas Kuhn said that a separate theory or a separate scientific achievement cannot master one’s mind so much as to fully set the development vector of a particular branch of knowledge; only a certain set of such theories can do this.

One of them can be refuted or modified, but in general, the structure of knowledge remains essentially the same, and accordingly the vector, which sets the direction of development, is not altered.

Exaggerating somewhat, we can say that the scientific paradigm is the basis of that ideology (more correctly to speak about the system of scientific views), which is shared by the overwhelming majority of scientists working in this field and which is officially codified with texts (theories, etc.) recognized in science as classical or canonical.

Therefore, the paradigm - that is what makes the scientific belief system (with all their variegation and diversity), some single entity; it also forms a well-defined conceptual apparatus, a certain discourse, which allows members of the scientific community to feel themselves as one, opposing themselves to others.

All this together creates a certain scientific tradition, as well as formal and informal institutions that are complementary to it (the latter include, among other things, the various unwritten rules that most scientists adhere to).

Let's pay attention to the following circumstance: the scientific paradigm exists and develops in many respects like a living organism; some of its carriers (specific scientists) leave this world (as individual cells of a biological organism can die), but the informational entity under consideration as a whole continues to exist.

Moreover, the paradigm as such is only indirectly related to the memory of an individual scientist; rather, its existence should be correlated with collective memory, more precisely with that fragment of the global communication environment that is associated with scientific research and development. There is also no doubt that the scientific paradigm is an object of a purely informational nature, and having a pronounced own behavior. The argument in favor of considering the paradigm as a special information entity with its own behavior is that the corresponding information is distributed, and here the analogy with the neural network is quite appropriate. The loss of a separate part of the communication network does not affect its functioning as a whole.

There is no doubt that in modern conditions the development of informational entities generated by communications in society undergoes pronounced transformations, which, first of all, is due to the development of telecommunication technologies. In this regard, many authors, for example [18],[19], speak about the formation of a new media realism. It is worth emphasizing that the parameters characterizing communications in society are subject to well-defined regularities, which are already revealed at this stage of research [20],[21]. In the quoted work it was shown that the distribution of users according to the number of friends in social online networks is described by a very specific formula, and one of its parameters retains a constant value for all cities, without exception, for which data was collected. This testifies in favor of the conclusion that the processes of self-organization in the communication space do not simply obey certain laws, but proceed in much the same way in different cities.

3. Conclusion

This implies the following most important circumstance of direct practical importance. Higher education in modern conditions does not de facto deal with individuals, as it was in the days of Jan Amos Comenius. It interacts with non-trivial informational entities, whose behavior, in particular, determines how effective the training will be. Simplifying, the student will easily learn the provisions that are fixed by the surrounding communication environment, while the assimilation of information that does not fall into this category will be extremely difficult. They will be de facto rejected as "irrelevant", "uninteresting" or "not related to everyday life". The misunderstanding of this circumstance, the attempt to overcome the numerous crisis phenomena in modern higher education within the framework of a paradigm that goes back to Comenius, only exacerbates negative trends. Moreover, such a state of affairs creates a real threat to the intellectual sovereignty of Kazakhstan and other states of the Eurasian Economic Community [16].

Considerations of this kind suggest that the methodological elaboration of issues one way or another connected with the concept of "intellect" is also important from the point of view of overcoming the crisis phenomena inherent in modern higher education. Modernization of higher education requires the transformation of basic approaches to the organization of higher education, complementary to the significant changes taking place in modern society. This, in turn, requires an adequate interpretation of the entire system of concepts, one way or another connected with the "information" category. The formulation of the principle of dialectic symmetry proposed in this work allows us to do this, overcoming the many contradictions that abounded in the previously proposed interpretations of the concept "information" and "intellect" (broader, social, collective consciousness).

References

[1] Nikitina E A 2014 Artificial Intelligence: Philosophy, Methodology, Innovations Philosophical Problems of Information Technologies and Cyberspace 2 pp 108-122
[2] Kholodnaya M A 1990 DOES INTELLIGENCE EXIST AS A MENTAL REALITY? Questions of psychology 5 pp 121-128
[3] Trufanov S N 2012 WHAT IS INTELLIGENCE? Applied psychology and psychoanalysis 4 p 3
[4] Dennett D C 1991 Consciousness explained Back Bay Books
[5] Suleimenov I E, Nurtazin A A, Gabrielyan O A, Shaltykova D B, Tasbulatova Z S and Panchenko S V 2017 Bureaucracy from the point of view of the theory of self-organization Educational resources and technology 2(19) pp 36-44
[6] Suleimenov I, Gabrielyan O, Mun G, Panchenko S, Amirzhan T and Suleimenova K 2014 Voting Procedure and Neural Networks. Int. J. on Communications 3 pp 16-20
[7] Suleimenov I, Panchenko S, Gabrielyan O, Pak I 2016 Voting procedures from the perspective of theory of neural networks Open Engineering 6(1)
[8] Gabrielyan O A, Mun G A, Panchenko S V and Suleimenov I E 2015 Dialectics of Information. Scientific notes Crimean Federal University named after V.I. Vernadsky. Philosophy. Political science. Culturology 2 pp 137-148
[9] Spirkin A G 2006 Philosophy: Textbook Moscow Gardariki
[10] Sociological Encyclopedic Dictionary Moscow 1998
[11] Satarov G A 2007 Public Opinion and Public Consciousness: Reality and Myth Social Sciences and the Present 4 pp 5-23
[12] Kirschenmann P K 2012 Information and Reflection: on Some Problems of Cybernetics and How Contemporary Dialectical Materialism Copes with Them Springer Science & Business Media
[13] Mamardashvili M K 1968 Analysis of consciousness in the works of Marx Questions of philosophy 6 pp 14-25
[14] Suleimenov I E and Grigoriev P E 2008 The physical bases of noospherology Almaty–Simferopol
[15] Suleimenov I, Massalimova A, Bakirov A and Gabrielyan O 2018 Neural Networks and the Philosophy of Dialectical Positivism MATEC Web Conf. 214
[16] Yegnanarayana B 2009 Artificial neural networks PHI Learning Pvt. Ltd
[17] Kuhn T S 1962 The Structure of Scientific Revolutions Chicago: University of Chicago Press
[18] Savchuk V V 2009 Medioreality. Media Subject. Media Philosophy. Media Philosophy II. The boundaries of the discipline 2 pp 226-241
[19] Kuznetsova E I 2009 Mediareality as a communicative medium Media Philosophy 2 pp 67-79
[20] Suleimenov I and Bakirov A 2018 Semi-empirical Theory of Distributions of Users of Social Online Networks by the Number of Contacts MATEC Web Conf. 214
[21] Kalimoldayev M N, Pak I T, Baipakbayeva S T, Mun G A, Shaltykova D B and Suleimenov I E 2018 Methodological basis for the development strategy of artificial intelligence systems in the Republic of Kazakhstan News of the national academy of sciences of the Republic of Kazakhstan-series of geology and technical sciences 6 pp 47-54.

IOP Conf. Series: Materials Science and Engineering **630** (2019) 012008 doi:10.1088/1757-899X/630/1/012008

Improving Energy Efficiency of Hybrid Beamforming System

Seulgi Lee[1], Won-Seok Lee[1], Seong-Joon Shim[1], Tae-Hoon Kang[1] and Hyoung-Kyu Song[1*]

[1]Department of Information and Communications Engineering, uT Communication Research Institute, Sejong University, Seoul 05006, Korea

Corresponding author's email : songhk@sejong.ac.kr[1*]

Abstract. Hybrid beamforming which combines conventional digital beamforming with analog beamforming system is promising model in wireless communication. The problem is using RF chain in digital precoder spends high cost and variable phase shifter exist at analog precoder processing network. Dealing with this problem, many architectures emerge to make lower hardware complexity and power consumption. This paper proposes low-resolution solution using phase over-sampling(POS) and switches(SW) on Radio Frequency(RF) chain to improve energy efficiency. We analyze proposed POS-SW hybrid beamforming system. In the simulation, the estimate is channel capacity in different signal-to-noise and energy efficiency during increasing number of RF chain on 3 dimension(3D) channel environment. It is shown that the proposed system has almost same channel capacity and high energy efficiency compared to the conventional system.

1. Introduction

Millimeter wave(mmWave) in wireless communication is a key feature in the fifth generation(5G) communication. When using high frequency band, size of required antenna is under 1cm that make array to building transmission system due to not enough output power. The main component of hybrid beamforming structure is Radio Frequency(RF) chain and phase shifters(PSs) [1]. RF chain converts analog to digital data to transmit in low frequency and PSs help controlling beam phase and amplitude to develop the shape of beam. Digital beamforming system uses multiple RF chains performing with each antenna that spend high cost. Whereas, analog beamforming system consists of only one RF chain and multiple PSs that have an antenna array. The main problem comes from the high-resolution phase constraints of the PSs. Hybrid beamforming has been proposed as a possible solution [2].

In mmWave communication system, conventional hybrid beamforming is required the large number of high-resolution PSs to construct analog beamformers to solve higher power consumption and hardware complexity. For limited technology of device, hybrid beamforming system has been studied to address the issue in low-resolution PS system. 3 dimension(3D) beamforming is used for fully dynamic adaptation of the beamforming. 3D beamforming based on 3D channel model that were designed to support antenna configurations that are capable of adaptation in azimuth only.

In this paper, the authors propose advanced hardware efficient phase over-sampling(POS)-switches(SW) system based on hybrid beamforming [3]. The limited number of POSs and SWs in analog beamformer part depends on input data bits and antenna array which achieves high energy efficiency as well as maintains high-resolution channel capacity [4]. Although POS-SW architecture is simple, antenna array steers proper beam shape in transmitter. This paper introduces conventional, proposed architecture and 3D channel model. Next, simulation result shows trade-off relationship between channel capacity and energy efficiency on 3D channel environment.

IOP Conf. Series: Materials Science and Engineering **630** (2019) 012008 doi:10.1088/1757-899X/630/1/012008

2. Hybrid beamforming

2.1. Traditional PS structure based on hybrid beamforming

'Figure 1' shows system model of conventional hybrid beamforming. At the transmitter, a baseband digital precoder process N_s input data stream to handle N_{RF} RF chain. Each RF chain combines N_{Tx} antenna array with PSs called analog precoder. $N_{Tx} \times N_{RF} = N_T$ antennas and N_T PSs are used in transmission stage.

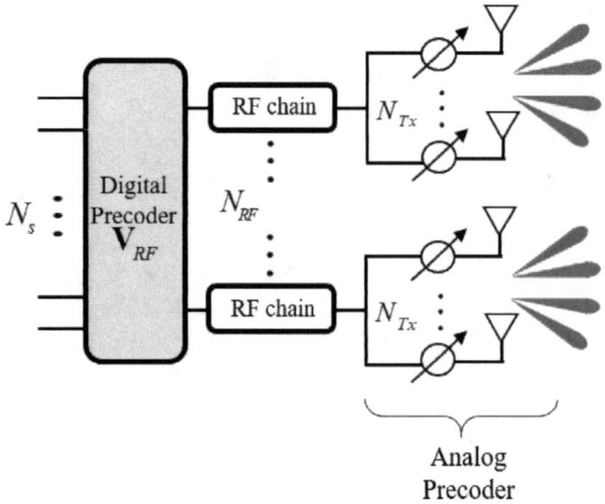

Figure 1. Traditional Hybrid Beamforming System

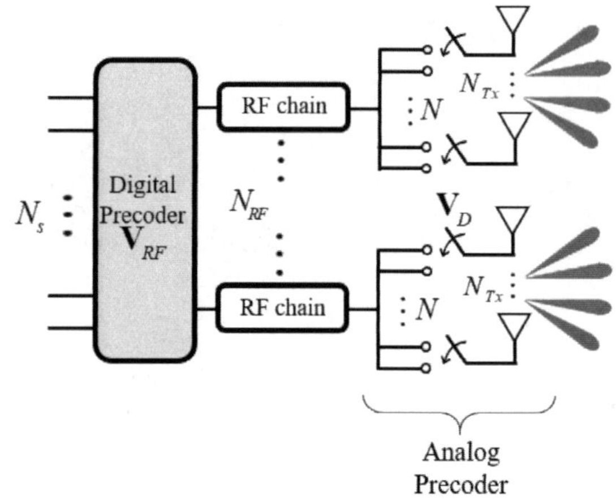

Figure 2 POS-SW Hybrid Beamforming System

The general equation of transmit signal for the K users is as follows:

$$\mathbf{x} = \mathbf{V}_{RF}\mathbf{V}_D\mathbf{s} = \sum_{l=1}^{K} \mathbf{V}_{RF}\mathbf{V}_{D_l}\mathbf{s}_l, \tag{1}$$

where $\mathbf{V}_{RF}$ represents digital precoder matrix and $\mathbf{V}_D$ represents analog precoder that dimension is $(N_T) \times N_{RF}$. Thus, the $k-$th received signal after decoding processing is as follows:

$$\mathbf{y}_k = \mathbf{H}_k\mathbf{V}_{RF}\mathbf{V}_{D_k}\mathbf{s}_k + \mathbf{H}_k\sum_{l \neq k}\mathbf{V}_{RF}\mathbf{V}_{D_l}\mathbf{s}_l + \mathbf{n}_k, \tag{2}$$

where $\mathbf{s}_k$ represents the $k-$th user data stream and $\mathbf{n}_k$ denotes Gaussian noise vector that expectation is 0 and variance is 1. Then, the final processed signal can be written as follows:

$$\mathbf{y}_k = \mathbf{W}_{RF}{}^H\mathbf{H}_k\mathbf{V}_{RF}\mathbf{V}_{D_k}\mathbf{s}_k + \mathbf{W}_{RF}{}^H\mathbf{H}_k\sum_{l \neq k}\mathbf{V}_{RF}\mathbf{V}_{D_l}\mathbf{s}_l + \mathbf{W}_{RF}{}^H\mathbf{n}_k. \tag{3}$$

In such a system, the overall $k-$th user channel capacity R_k can be written as follows [5]:

$$R_k = \log_2\left\{1 + \mathbf{W}_{RF}\mathbf{C}_k{}^{-1}\mathbf{W}_{RF}{}^H\mathbf{H}_k\left(\mathbf{V}_{RF}\mathbf{V}_{D_k}\right)\left(\mathbf{V}_{RF}\mathbf{V}_{D_k}\right)^H\mathbf{H}_k{}^H\right\}. \tag{4}$$

IOP Conf. Series: Materials Science and Engineering **630** (2019) 012008 doi:10.1088/1757-899X/630/1/012008

The covariance of the interference plus noise at user k $\mathbf{C}_k$ is as follows:

$$\mathbf{C}_k = \mathbf{W}_{RF}{}^H \mathbf{H}_k \left\{ \sum_{l \neq k}^{K} \left(\mathbf{V}_{RF} \mathbf{V}_{D_l} \right) \left(\mathbf{V}_{RF} \mathbf{V}_{D_l} \right)^H \right\} \mathbf{H}_k \mathbf{W}_{RF} + \sigma^2 \mathbf{W}_{RF}{}^H \mathbf{W}_{RF}. \tag{5}$$

Based on the energy consumption, since N_{RF} RF chain and N_T PSs are used in system model, the energy efficiency η can be designed as follows [6]:

$$\eta_{N=\infty} = \frac{R_k}{P_t + N_{RF} P_{RF} + N_T P_{PS}} (\text{bps/Hz/W}), \tag{6}$$

where P_t is the transmitted energy, P_{RF} is the energy consumption by RF chain, P_{PS} is the energy consumed by PS.

2.2. Proposed system architecture : POS-SW

'Figure 2' shows system model of proposed hybrid beamforming. The system utilizes the limited number of simple POSs and switches network. The number of POSs is decided by the resolution of phase. Resolution of analog precoder is N and the number of POSs is 2^N. POSs values are complex plane angles between maximum and minimum in each antenna array phase divided by 2^N. POSs values $v_{\mathbf{V}_{Tx_{RF}},n}$ express as follows:

$$v_{\mathbf{V}_{Tx_{RF}},n} = \min\left(\angle \mathbf{V}_{Tx_{RF}} \right) + \left(\frac{\max\left(\angle \mathbf{V}_{Tx_{RF}} \right) - \min\left(\angle \mathbf{V}_{Tx_{RF}} \right)}{2^N - 1} \right) (n-1), \left(n = 1, 2, \cdots, 2^N \right), \tag{7}$$

where $\angle \mathbf{V}_{Tx_{RF}}$ is the matrix of phase analog precoder elements at $Tx-$th RF chain.

Since POSs are connected at antenna array, the number of switches is same as N_{Tx}. The proposed hybrid beamforming system also uses digital and analog precoder in different way. Then, the channel capacity is same as conventional system like equation (4). The proposed system energy efficiency is different because conventional PS are replaced by SW. POSs are implemented by simpler inverters or microstrip delay lines. So, POSs have negligible power consumption. The energy efficiency η can be redesigned as follows:

$$\eta_N = \frac{R_k}{P_t + N_{RF} P_{RF} + N_T P_{SW}} (\text{bps/Hz/W}), \tag{8}$$

where P_{SW} is the energy consumed by switch.

2.3. 3D channel model

3D channel model is indispensable system for beamforming system to shape beam pattern. This section describes the 3D channel model proposed by the 3rd generation partnership project(3GPP) and how to express analog precoder based on proposed POS-SW system. Designing hybrid precoder, digital and analog precoder, requires full knowledge of channel state information(CSI). CSI is shown in 3D geometric stochastic model, describing the scattering environment between base station and users. To design 3D channel, the angle of shape of beam direction expressed by θ the zenith direction and ϕ the azimuth direction [7]. The r_{rx} and r_{tx} are each receiver and transmitter aspherical unit vectors expressed in Cartesian coordinates. They are defined as follows:

$$r_{rx}, r_{tx} = \begin{pmatrix} x = \sin\theta\cos\phi \\ y = \sin\theta\sin\phi \\ z = \cos\theta \end{pmatrix}. \quad (9)$$

The $\bar{d}_{BS}$ is the location vector of basement station antenna array element, the $\bar{d}_{BS}$ is as follows:

$$\bar{d}_{BS} = \begin{pmatrix} 0 \\ (p-1)d_H \\ (q-1)d_V \end{pmatrix}, \quad (10)$$

where the p and q are index number of antenna element in the horizontal and vertical direction. The d_H is element spacing in the horizonal direction and d_V is element spacing the vertical direction. In hybrid beamforming system, using r_{rx} and antenna array is important point to design analog precoder.

3. Simulation Result

In this section, simulation results are presented to show the difference performance between the conventional and proposed system. As simulation parameter, the number of data streams is 4 and $N_{RF} = 4$, $N_{Tx} = 4$. The users are $K = 2$ and the channel path is 7 in full dimension fading channel. Based on orthogonal frequency division multiplexing(OFDM) system, FFT size is 64 and cyclic prefix(CP) size is 16.

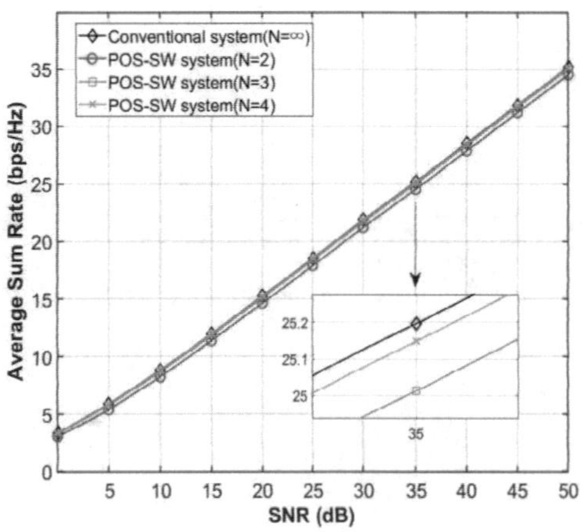

Figure 3. Channel capacity comparison of conventional and POS-SW system

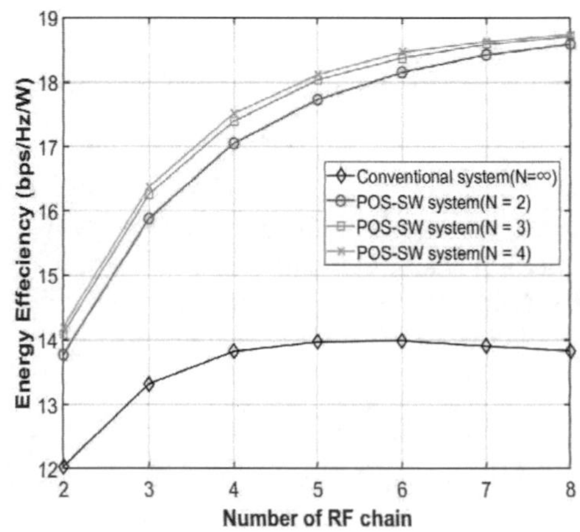

Figure 4. Energy efficiency comparison of conventional and POS-SW system

'Figure 3' shows the different channel capacity for conventional and proposed system. Channel capacity of conventional system is the highest because of infinite resolution $N = \infty$ which means PSs steer proper direction in analog beamformer transmitter. In contrast, using POSs and SW has lower channel capacity than conventional system because of low-resolution. 'Figure 3' proves that high resolution has similar channel capacity as infinite resolution. In addition, the gap between each line goes lower having high resolution.

'Figure 4' shows energy efficiency POS-SW system and conventional system in terms of the number of RF chains based on equation (6), (8). The energy efficiency is defined as the ratio of spectral efficiency to the total power consumed at the transmitter. The simulation uses general power values of SW and PS, P_{SW} is 5mW and P_{PS} is 40mW [4]. Transmitted energy P_t assumed 1W. Signal-to-noise ratio is 30dB, as expected, conventional system model is lower estimate than POS-SW system and using 16($N=4$) POSs system has best energy efficiency. The reason of this difference is PSs consume high power consumption than phase over sampler. The result of energy efficiency is trade-off relation with channel capacity. It can be observed that the proposed POS-SW system gives significant energy efficiency advantages comparing to conventional infinite resolution hybrid system.

4. Conclusion

This paper analyses the POS-SW structure on hybrid beamforming system for mmWave communication on 3D channel model. The limited number of POS decides on the resolution of antenna array and the value of POSs are designed from maximum and minimum phase angle antenna array. Energy efficiency goes higher without using phase shifter although channel capacity is nearly similar as conventional system. All system utilized 3D channel model that considering steering angle of beam in CSI. Finally, simulation result shows that proposed system has almost same performance of channel capacity and the higher energy efficiency compared to the conventional hybrid beamforming system. For future research, it would be interesting to investigate that POS-SW model is same performance of full digital beamforming capacity spend less energy.

5. Acknowledgments

This work was supported by Institute for Information & communications Technology Promotion(IITP) grant funded by the Korea government(MSIT) (No.2017-0-00217, Development of Immersive Signage Based on Variable Transparency and Multiple Layers) and was supported by the MSIT(Ministry of Science and ICT), Korea, under the ITRC(Information Technology Research Center) support program(IITP-2019-2018-0-01423) supervised by the IITP(Institute for Information & communications Technology Promotion). *Corresponding author: Hyoung-Kyu Song

Reference

[1] Andreas F M, Vishnu V, Shengqian H, Zheda L, Sinh L H N, Linsheng L and Katsutuki H. 2017. Hybrid Beamforming for Massive MIMO: A survey. *IEEE Communications Magazine.* **55**. pp.134-41

[2] Robert W H J. Nuria G. Sundeep R. Wonil R and Akbar M S. 2016. A Overview of signal processing techniques for millimetre wave MIMO system. *IEEE Wireless Communications.* **10(3)**. pp.436-53.

[3] Ming L, Zihuang W, Hongyu L, Qian L and Liang Z. 2019. A Hardware-efficient hybrid beamforming solution for mmWave MIMO systems. *IEEE Wireless Communications.* **1**. pp.137-43

[4] Roi M, Cristian R, Nuria G, Ahmed A, Robert W H. 2016. Hybrid MIMO architectures for millimeter wave communications: phase shifters or switches?. *IEEE Wireless Communications.* **4**. pp.247-67

[5] Foad S and Wei Y. 2016. Hybrid digital and analog beamforming design for large-scale antenna arrays. *IEEE Wireless Communications.* **10(3)**. pp.501-13

[6] Xinyu G, Linglong D, Shuangfeng H, Chin-Lin I and Robert W H. 2016. Energy-efficient hybrid analog and digital precoding for mmWave MIMO systems with large antenna arrays. *IEEE Wireless Communications.* **34(4)**. pp.998-1009

[7] Fjolla A, Martin T and Markus R. 2016. 3GPP 3D MIMO channel model: A holistic implementation guideline for open source simulation tools. *EURASIP Journal on Wireless Communications and Networking.* **2016(55)**.

Antenna Diversity Scheme for Broadcasting System Signal Detection

Tae-Hoon Kang[1], Seulgi Lee[1], Jae-Hyun Ro[1] and Hyoung-Kyu Song[1*]

[1]Department of Information and Communications Engineering, uT Communication Research Institute, Sejong University, Seoul 05006, Korea

Corresponding author's e-mail address : songhk@sejong.ac.kr[1*]

Abstract. In order to improve robustness of bootstrap signal, this paper studies on multi-antenna transmit and receive diversity. This paper proposes bootstrap symbol detection scheme using multi-antenna technique for ATSC 3.0 system to improve detection performance. Multi-antenna technique has diversity scheme for obtaining SNR gain and multiplexing scheme for improving data rate. This paper treats transmit and receive diversity scheme for detecting bootstrap symbol by space time block code(STBC) and maximum ratio combining(MRC). The simulation results show that the performance of the proposed scheme is improved when the first bootstrap symbol at a symbol error rate of 0.01 obtains a signal to noise ratio(SNR) gain of 4dB compared to 1 by 1 single-input single-output(SISO).

1. Introduction

Advanced television system committee(ATSC) is standardization group for digital television in USA. Recently, ATSC developed standard called ATSC 3.0 and this standard has been adopted at Korea in 2016. This standard constitutes the normative specification for the initial entry point of a physical layer waveform [1][2]. The A/321 standard was established in March 2016 and the standard for bootstrap symbols was published in this document.

In this standard, bootstrap uses several fixed configuration (e.g., sampling rate, signal bandwidth, subcarrier space, time domain structure). The bootstrap is also located in front of the physical layer frame and transmits a total of four symbols in the ATSC 3.0 specification. Bootstrap symbol is generated by multiplying Zadoff-Chu sequence and pseudo-noise sequence in the frequency domain. In time domain, each symbol except the first symbol has a cyclic shifted structure and represents symbol information according to the degree of cyclic shift relative to adjacent symbols [3].

This paper deals with antenna diversity technique, which is used in communication systems, to improve robustness of bootstrap symbols. The spatial diversity scheme can be classified into a transmit and receive diversity scheme. In this paper, a space-time block code(STBC) of Alamouti is used as a transmit diversity and a maximum ratio combining(MRC) is used as a receive diversity. STBC is a technique to encode transmission signals using multiple transmit antennas and maximizes diversity gain by using orthogonality [4]. Similar to STBC, MRC is a technique to maximize diversity gain using multiple receive antenna.

2. System model

2.1. Structure of bootstrap signals

IOP Conf. Series: Materials Science and Engineering **630** (2019) 012009 doi:10.1088/1757-899X/630/1/012009

In frequency domain, bootstrap symbols is generated by multiplying Zadoff-Chu(ZC) sequence and pseudo-noise(PN) sequence. The ZC sequence is as follows:

$$z_q(k) = e^{-j\pi q \frac{k(k+1)}{N_{ZC}}}, \, k = 0,1,2,\cdots,N_{zc}-1, \quad (1)$$

where k is subcarrier number, q is the root of the ZC sequence which means the major version of the broadcasting signal and $N_{zc} = 1499$ is the number that results in a channel bandwidth $4.5MHz$ with a subcarrier spacing of $f_\Delta = 3kHz$ except DC component. For ATSC 3.0, root of ZC sequence is 137 which means that major version is 0.

PN sequence is generated by linear feedback shift register(LFSR) with length $l = 16$. PN sequence generator polynomial is given by $p(x) = x^{16} + x^{15} + x^{14} + x + 1$. Initial state of PN sequence generator, called seed, indicates the minor version. The sequence generated by multiplying ZC sequence and PN sequence forms a symmetry around the DC component. Each symbol has a total of 2048 subcarriers, but uses only 1498 subcarriers excluding the DC component. The formula of the generated sequence is as follows:

$$s_n(k) = \begin{cases} z_q(k+749)c(749(n+1)+k) & , -749 \le k < 0 \\ z_q(k+749)c(749(n+1)-k) & , 1 \le k < 749 \\ 0 & , otherwise \end{cases}, \quad (2)$$

where z_q, c denote ZC and PN sequence respectively, and $s_n(k)$ denotes the $k-$th sequence in the $n-$th symbol.

$s_n(k)$ is transformed into a time domain signal with 2048 sampling length, and the cyclic shift of the signal is performed. Cyclic shift has relative, absolute cyclic shift. The cyclic shift is represented by the cyclic shift value of the current symbol relative to the previous symbol and the absolute cyclic shift is expressed as the sum of the previous symbol and the current symbol. When the cyclic shifted symbol is $A_n(l)$, n and l mean symbol and sample number, respectively. Finally, the bootstrap symbol consists of a combination of three areas: A, B and C in the time domain. The first bootstrap symbol is generated with the structure shown in 'figure 1'. The remaining bootstrap symbols are generated into the structure shown in 'figure 2'.

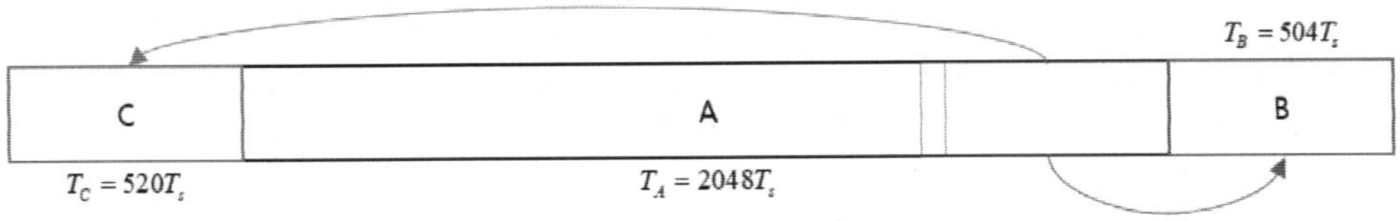

Figure 1 CAB time domain symbol structure for first bootstrap

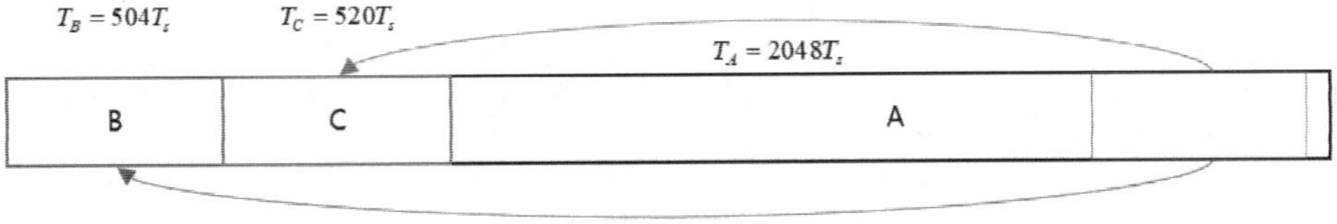

Figure 2 BCA time domain symbol structure for the remaining bootstraps

2.2. *Proposed spatial diversity scheme for bootstrap*

First, the block shape of the transmission signal of the 2×2 STBC, which is a transmit diversity technique, is as follows:

$$\mathbf{S} = \begin{bmatrix} s_0 & s_1 \\ -s_1^* & s_0^* \end{bmatrix}, \tag{3}$$

where each row represents a time slot and each column represents a transmit antenna. Assuming a quasi-static channel between transmission time slots, the STBC receive signal is as follows:

$$\mathbf{Y}_s = \begin{bmatrix} \mathbf{y}_{t_1} \\ \mathbf{y}_{t_2} \end{bmatrix} = \begin{bmatrix} y_{t_1,1} \\ y_{t_1,2} \\ y_{t_2,1} \\ y_{t_2,2} \end{bmatrix} = \begin{bmatrix} h_{1,1}s_0 + h_{1,2}s_1 \\ h_{2,1}s_0 + h_{2,2}s_1 \\ -h_{1,1}s_1^* + h_{1,2}s_0^* \\ -h_{2,1}s_1^* + h_{2,2}s_0^* \end{bmatrix} + \begin{bmatrix} w_{t_1,1} \\ w_{t1,2} \\ w_{t_2,1} \\ w_{t_2,2} \end{bmatrix}, \tag{4}$$

where $\mathbf{y}_{t_n}$ $(n = 1,2)$ is received symbol in the n th timeslot, $y_{t_n,j}$ is the received signal at the $j-$th receiver in the n th time slot, $h_{i,j}$ $(i = 1,2,\ j = 1,2)$ is the channel coefficient between the $j-$th receiver and the $i-$th transmitter and $w_{t_n,j}$ is additive white Gaussian noise(AWGN) in the $n-$th timeslot at the $j-$th receiver. In addition, the STBC receiver can detect the transmitted signal in a simple linear combination due to the orthogonality [5]. The detection equation is as follows:

$$\hat{\mathbf{S}} = \begin{bmatrix} s_0 & s_1 \end{bmatrix} = \begin{bmatrix} \dfrac{\sum_{j=1}^{N_r} h_{j,1}^* y_{t_1,j} + h_{j,2} y_{t_2,j}^*}{\sum_{j=1}^{N_r}\sum_{i=1}^{2} \left|h_{i,j}\right|^2} & \dfrac{\sum_{j=1}^{N_r} h_{j,2}^* y_{t_1,j} - h_{j,1} y_{t_2,j}^*}{\sum_{j=1}^{N_r}\sum_{i=1}^{2} \left|h_{i,j}\right|^2} \end{bmatrix}. \tag{5}$$

Similar to STBC, receive diversity is a technique that maximizes diversity gain using multiple receive antennas. In the $N_t \times N_r$ system, the equation representing the MRC received signal is as follows:

$$\mathbf{Y}_M = \begin{bmatrix} y_1 \\ y_2 \end{bmatrix} = \begin{bmatrix} h_{1,1} & h_{1,2} \\ h_{2,1} & h_{2,2} \end{bmatrix} \mathbf{s} + \begin{bmatrix} w_1 \\ w_2 \end{bmatrix}, \tag{6}$$

where y_i is the receive signal at the $i-$th receiver, $\mathbf{s}$ is transmit signal and w_i is the $i-$th AWGN at receiver. The MRC scheme achieves diversity gain by performing a linear combination operation. The transmit signal detection equation of the MRC scheme is as follow:

$$\hat{s} = \frac{\left(h_{1,1} + h_{2,1} + h_{1,2} + h_{2,2}\right)^* y_i}{\left|h_{1,1} + h_{2,1} + h_{1,2} + h_{2,2}\right|^2}. \tag{7}$$

In this paper, STBC and MRC are used to improve detection performance. Bootstrap symbol structure for using STBC is shown ‘figure 3’.

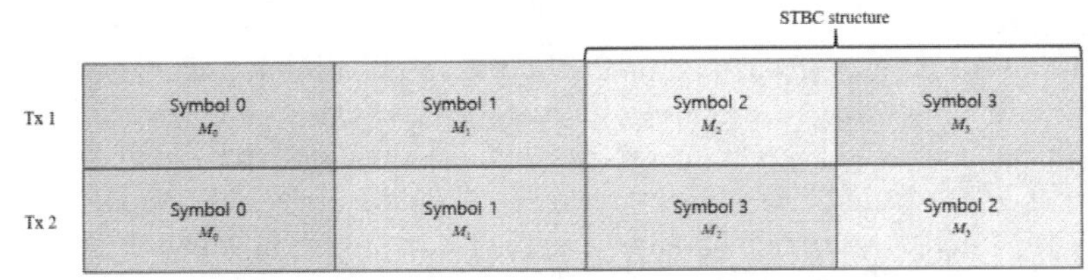

Figure 3 transmitted bootstrap structure with STBC

In addition, in order to make the STBC structure, the cyclic shift effect from the structure of the bootstrap signal generator should be processed in the frequency domain unlike the conventional one.

$$IFFT\{x((k-M) \bmod 2048)\} = \mathrm{X(k)e}^{\left(j\frac{2\pi kM}{2048}\right)} \tag{8}$$

The modified bootstrap structure consists of four kinds of orthogonal frequency division multiplexing(OFDM) symbols in the bootstrap frame and M_n $(n=0,1,2,3)$ is absolute cyclic shift value applied OFDM symbol. In this symbol, each column is timeslot and $M_0=0$. STBC structure is only applied in timeslot 3 and 4, since symbol 0 uses the CAB structure and the remaining symbols use the BCA structure in 'figure 1' and 'figure 2'. In the 2×2 system, received signal is as follow:

$$\mathbf{Y}_s = \begin{bmatrix} \mathbf{y}_0 \\ \mathbf{y}_1 \\ \mathbf{y}_2 \\ \mathbf{y}_3 \end{bmatrix} = \begin{bmatrix} y_{1,}^0 & y_2^0 \\ y_{1,}^1 & y_2^1 \\ y_1^2 & y_2^2 \\ y_1^3 & y_2^3 \end{bmatrix} = \begin{bmatrix} (h_{1,1}+h_{1,2})s_0 & (h_{2,1}+h_{2,2})s_0 \\ (h_{1,1}+h_{1,2})s_1 & (h_{2,1}+h_{2,2})s_1 \\ (h'_{1,1}+h'_{1,2})s_2 & (h'_{2,1}+h'_{2,2})s_3 \\ -(h'_{1,1}+h'_{1,2})s_3^* & (h'_{2,1}+h'_{2,2})s_2^* \end{bmatrix} + \begin{bmatrix} w_1^0 & w_2^0 \\ w_1^1 & w_2^1 \\ w_1^2 & w_2^2 \\ w_1^3 & w_2^3 \end{bmatrix}, \tag{9}$$

where y_k^n is received signal at the $k-$th receiver in the $n-$th timeslot, $h_{k,j}^n$ is the channel coefficient between the $j-$th receiver and the $i-$th transmitter in the $n-$th timeslot, s_m is bootstrap symbol and w_k^n is AWGN. Also, assuming a quasi-static channel environment during two time slots, timeslot 1 has the same channel coefficients as 2, and timeslot 3 has the same channel coefficients as 4. To obtain cyclic shift $\tilde{M}_1$, the received signal combining channel coefficient $(h_{1,1}+h_{1,2})$ is used.

$$\hat{s}_1 = \frac{\left(h_{1,1}+h_{2,1}+h_{1,2}+h_{2,2}\right)^* y_1}{\left|h_{1,1}+h_{2,1}+h_{1,2}+h_{2,2}\right|^2}. \tag{10}$$

Also, STBC demodulation is performed by $\mathbf{y}_2$ and $\mathbf{y}_3$ to obtain $\tilde{M}_2$ and $\tilde{M}_3$ as follows:

$$\begin{bmatrix} \hat{s}_2 \\ \hat{s}_3 \end{bmatrix} = \begin{bmatrix} \dfrac{(h'_{1,1})^* y_1^2 + h'_{1,2}(y_1^3)^* + (h'_{2,1})^* y_2^2 + h'_{2,2}(y_2^3)^*}{|h'_{1,1}|^2+|h'_{1,2}|^2+|h'_{2,1}|^2+|h'_{2,2}|^2} \\ \dfrac{(h'_{1,2})^* y_1^2 - h'_{1,1}(y_1^3)^* + (h'_{2,2})^* y_2^2 + h'_{2,1}(y_2^3)^*}{|h'_{1,2}|^2+|h'_{1,1}|^2+|h'_{2,2}|^2+|h'_{2,1}|^2} \end{bmatrix}. \tag{11}$$

Since the receiver knows the value of the bootstrap symbol that is not cyclic shifted, the value of absolute cyclic shifted M_n is known by multiplying the detected symbol by the conjugate of the symbol that has is not cyclically shifted.

3. Simulation results

'Figure 4' shows the SER performance of the bootstrap transmission frame of the proposed scheme and the conventional scheme in the 7-tap multi-path Rayleigh fading environment. When multiple transmit antennas are used, the transmit power is normalized for comparison under the same conditions as the 1x1 fading channel environment. Symbol 1 obtains a SNR gain of about $4dB$ in 2×2 system when the SER is 10^{-3} compared to symbol 1 in SISO system.

'Figure 5' shows the SER performance results of proposed scheme according to the number of receiving antennas in the 7-tap multi-path Rayleigh fading environment. In the 2×1 system, the receive diversity gain is not obtained and the performance is worse than 2×2. In addition, the 2×3 system has better diversity gain than 2×2 and improved performance.

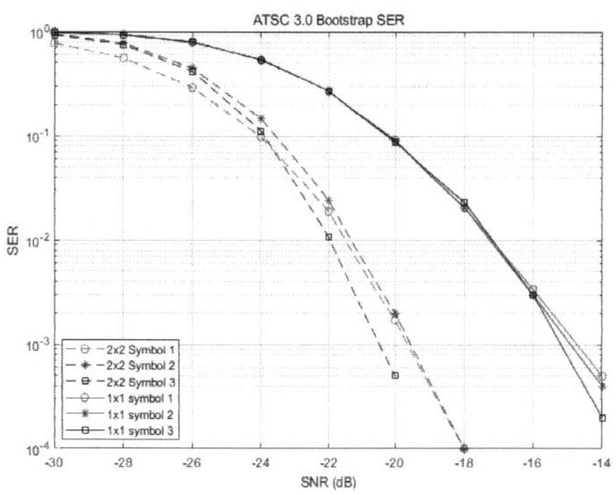

Figure 4 The SER performance of the bootstrap transmission frame of the proposed scheme and the conventional scheme

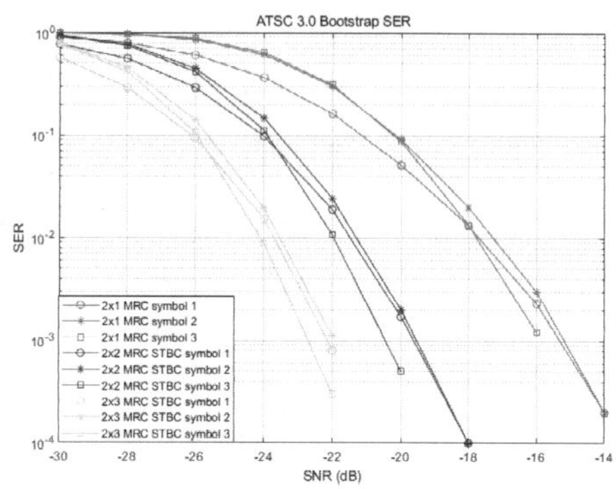

Figure 5 SER performance results of proposed scheme according to the number of receiving antennas

4. Conclusion

This paper shows the structure and generation process of the bootstrap symbol of the ATSC 3.0 standard, and then proposes a technique for obtaining diversity gain using STBC and MRC, which are transmit and receive diversity techniques. The proposed scheme improves detection performance by using multiple antennas to obtain diversity gain. The simulation results show that 2×2 multi antenna diversity scheme can improve SER performance compared to the 1×1 system under the Rayleigh fading environments. The increase in the number of transmit and receive antennas shows a performance improvement in the proposed scheme.

5. Reference

[1] ATSC. 2016. System Discovery and Signaling, document A/321. (Washington D.C. : Adv TV Syst Committee).
[2] L. Fay, L. Michael, D. Gómez-Barquero, N. Ammar, and M. W. Caldwell. 2016. An overview of the ATSC 3.0 physical layer specification. *IEEE Trans. Broadcast.* **62** (1). pp.159–71
[3] Kim H S, Kim j C, Park S I, Hur N H, Simon M and Aitken M. 2018. A Novel Iterative Detection Scheme of Bootstrap Signals for ATSC 3.0 System. IEEE Trans. Broadcast.
[4] Alamouti S M. 1998. A simple transmit diversity technique for wireless communications. IEEE J.Select Areas Commun. **l6** (8). pp.1451-58
[5] Vucetic B and Yuan J. 2003. Space-Time Coding. (New York: Wiley)

IOP Conf. Series: Materials Science and Engineering **630** (2019) 012010 doi:10.1088/1757-899X/630/1/012010

Design of Home Intelligent Alarm System Abstract

Bofeng Ding[1], Chengliang Zhang[1], Lei Li[2], Kang Sun[3] and Hongbin Dou[1]

[1]School of Mechanical Engineering, University of Jinan, Jinan 250022, China
[2]School of Mechanical and Automotive Engineering, Qilu University of Technology (Shandong Academy of Sciences), Jinan 250353, China
[3]Jinan Urban Construction Group Co., Ltd. Jinan 250031, China

me_zhangcl@ujn.edu.cn

Abstract. This paper describes the research and development of home intelligent emergency alarm system and its working principle. In the design, STC12C5A60S2 single chip microcomputer is used as the control core of the alarm system, which controls infrared sensor, MQ-2 smoke sensor, SIM900A module, buzzer, LCD1602 display screen and GSM module. The corresponding circuit of each module is designed, and each part of the circuit is integrated to complete the required function.

1. Introduction

With the development of computer technology, especially single-chip microcomputer technology, single-chip microcomputer has been more and more widely used in intelligent instruments, industrial control, daily life and many other fields. Family intelligent emergency alarm system can ensure the safety of residents' own lives and property. in the event of emergencies, it can inform the residents in a timely manner so as to reduce people's economic losses, and even save human lives [1]. Xinhua Zhao of Harbin University of Engineering has studied the infrared alarm system and realized the function of field alarm [2], Bo Sun of Shandong University of Science and Technology has studied the smoke alarm system and realized the high sensitivity detection result [3]. There is also the application of GSM module, but also shows its excellent adaptability. In this paper, through the integration and connection of infrared module, smoke alarm module and GSM module, a kind of home intelligent alarm system equipment with many functions is completed.

2. Design scheme

2.1. Basic design idea

This paper analyzes the practicality and design of home intelligent emergency call system at home and abroad, combined with the situation of Chinese residents and various aspects. The selection of components must meet the design requirements and be able to be used properly. Therefore, the system requires the following conditions:

- there should be human infrared sensors, which can detect intruders;
- there should be smoke sensors to detect gas leakage and fire;
- there should be an emergency call button, which can call the police at a critical moment;
- there should be a liquid crystal display (LCD), which can understand whether the system is normal and the sensor monitoring data;
- there should be GSM module, in order to ensure that residents know the accident in a timely manner, make the fastest response, send text messages to users.

2.2. *Overview of the overall programme*

This design is controlled by STC12C5A60S2 single chip microcomputer [4]. The fire alarm part adopts MQ2 smoke sensor to monitor [5], the anti-theft alarm part uses the human body infrared sensor [6], the emergency call part is designed in the form of keys, and the display part uses LCD1602 liquid crystal display. The alarm part is composed of buzzer and GSM module [7]. The GSM module sends an alarm text message to the household at the time of alarm. The design block diagram is shown in figure 1.

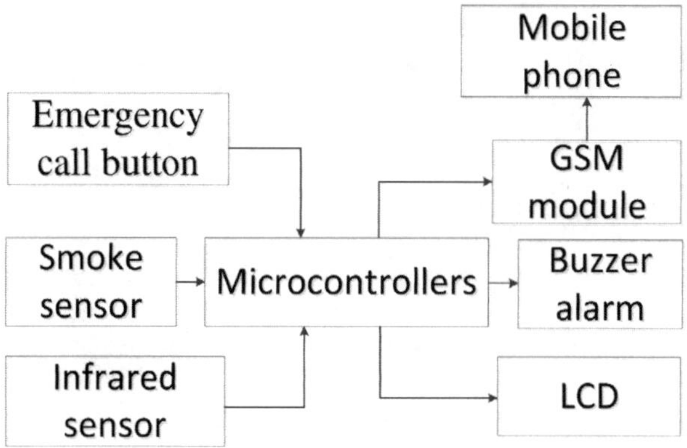

Figure 1. System principle block diagram

3. Specific Design of Family Emergency alarm system

3.1 STC12C5A60S2 microcontroller system

The home emergency alarm system automatically alarms through the system designed by the STC12C5A60S2 single-chip microcomputer. The console is connected to all modules, which is relatively stable in technology and relatively inexpensive. STC12C5A60S2 series MCU is a single-chip microcomputer of machine cycle (1T) produced by Hongjing Technology. It is a new generation of 8051 microcontrollers with high speed/low power/super interference immunity, and the instruction code is fully compatible with the traditional 8051. It integrates MAX810 dedicated reset circuit, 2-channel PWM, 8-channel high-speed 10-bit A/D conversion (250K/S) for motor control and strong interference. It is widely used in industrial automation control, data acquisition, intelligence, and mechatronics research.

3.2 Infrared sensor selection

The human infrared sensing system is based on the model HC-SR501. 8 ~ 14 μ m is the detectable wavelength, and the infrared wavelength emitted by human body is about 10 μ m, which is within the detection range. HC-SR501 sensors can be monitored 24 hours a day. When an illegal person enters, the sensor outputs a high level. after the illegal person is far away from the monitoring, the system closes the high level after delay and outputs a low level.

Figure 2. HC-SR501

3.3 Smoke sensor selection

MQ-2 for smoke sensors. MQ-2 sensor belongs to semiconductor smoke sensor, which has the advantages of good anti-interference, long service time, high sensitivity, large monitoring area and relatively simple circuit. Analog signal is the form of signal received by MQ-2 sensor, and A/D module can carry out analog-to-digital conversion. When the suspicious gas or smoke is monitored and increased to a certain value, the MQ-2 sensor will pass the signal processing to the single-chip microcomputer after AD, and finally the single-chip microcomputer controls the alarm. The MQ-2 sensor is buffered for a period of time before it works, and the accuracy is higher.

Figure 3. MQ2 smoke sensor **Figure 4.** ADC0832

ADC0832 converter is produced by NS (National Semiconductor) Company, with 8-bit serial interface, in which the working consumption is low, the performance is sufficient, the price is low, and it is widely used in many small machines.

3.4 GSM module selection

The type of GSM communication module designed in this paper is SIM900A. The GSM uses advanced technology to consume a particularly low current consumption of only 1 milliampere at the lowest level, and the interface is an industry-standard interface [8]. The main functional features of SIM900A are as follows:

- many functions: telephone, short message, GPRS network communication and so on;
- there are many power supply modes: USB charger and lithium battery can supply power;
- the protection function is complete: the power supply has reverse protection and SIM card protection.

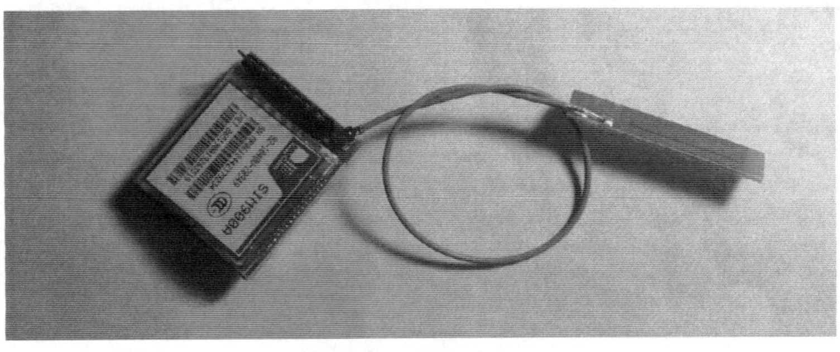

Figure 5. SIM900A

IOP Conf. Series: Materials Science and Engineering **630** (2019) 012010 doi:10.1088/1757-899X/630/1/012010

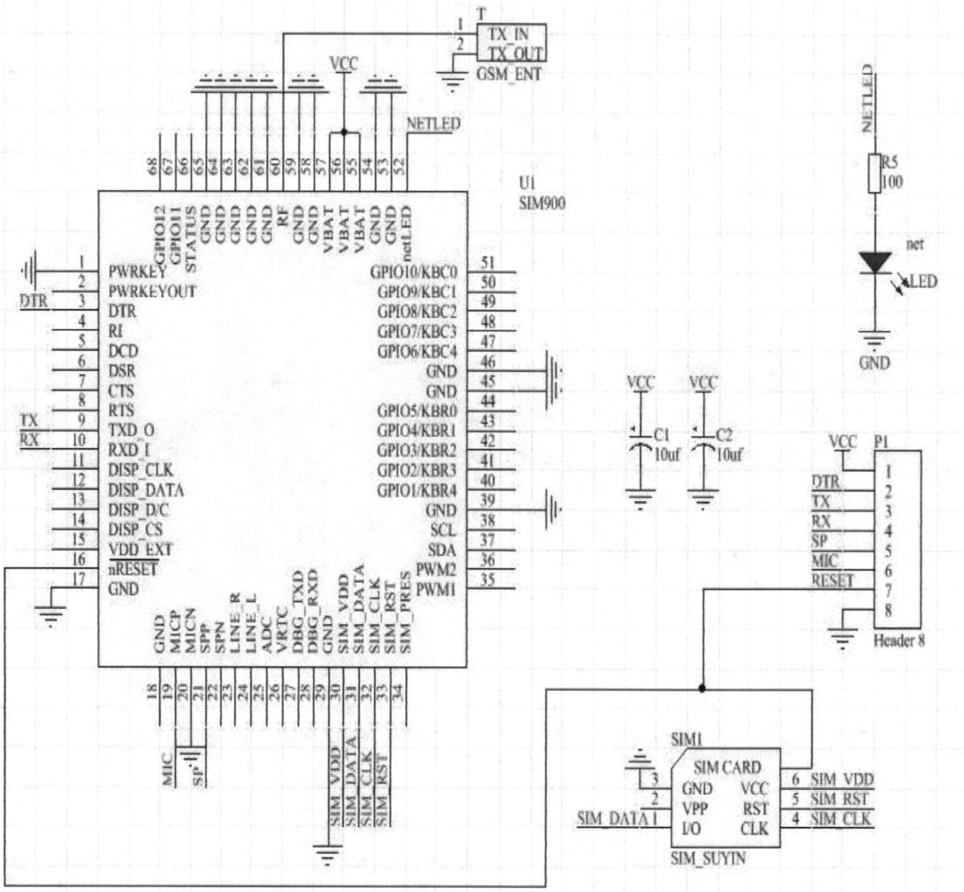

Figure 6. SIM900A schematic diagram

4. Hardware circuit design

4.1. Design of burglar alarm Circuit

The anti-theft module adopts infrared sensor. In order to make the alarm system safe and reliable, when illegal personnel break in, the infrared sensor can be found in time, and the infrared sensor can work with other sensors and key modules to protect the safety of residents. The infrared sensor consists of three pins, the first pin is connected to the power supply, the second pin is connected to the single-chip microcomputer, and the pull-up resistance is connected with the first pin, because the ability of this sensor to output high level is weak, and the third pin is grounded. The circuit diagram is shown in figure 7.

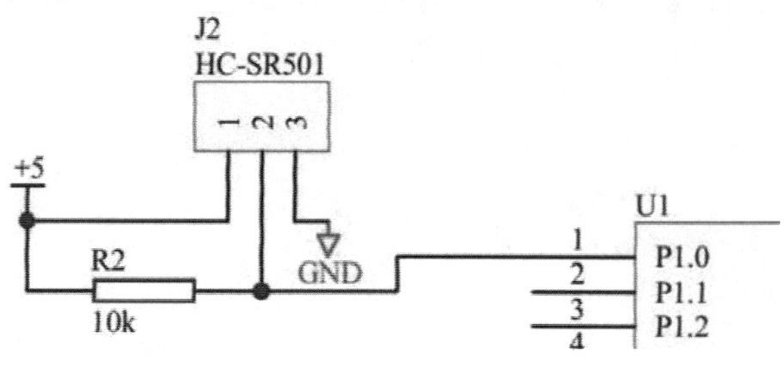

Figure 7. Burglar alarm circuit

4.2. *Fire alarm circuit design*

Home intelligent emergency alarm systems can also detect fires. The smoke sensor is selected, which can not only detect the fire, but also detect the gas leakage. The smoke sensor module model is MQ-2, to convert analog signal into digital signal through ADC0832 and then give it to STC12C5A60S2. The MQ-2 smoke sensor has six pins, pin 1.2.3 is connected to the power supply, pin 4.6 is connected to the AD module, and pin 5 is grounded. The circuit diagram of the MQ-2 smoke sensor is shown in figure 8.

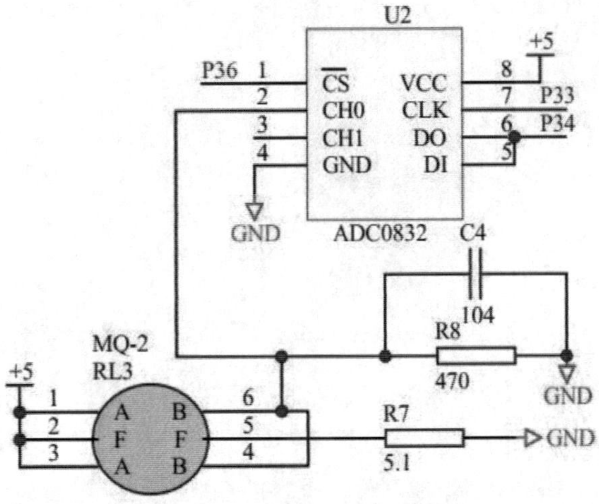

Figure 8. Fire alarm circuit

4.3. *Circuit Design of GSM Module*

When the sensor of the system receives the signal, in order to inform the residents in time, the wireless alarm function is designed, and the short message is sent to the residents through the wireless module. The model of the GSM module designed in this paper is SIM900A. STC12C5A60S2 sends text messages by controlling the GSM module. The TX of single chip microcomputer is connected to the RX of GSM, and the RX of single chip microcomputer is connected to the TX of GSM, so as to realize the receiving and sending of data, and the GND is connected to GND. The GSM interface circuit is shown in figure 9.

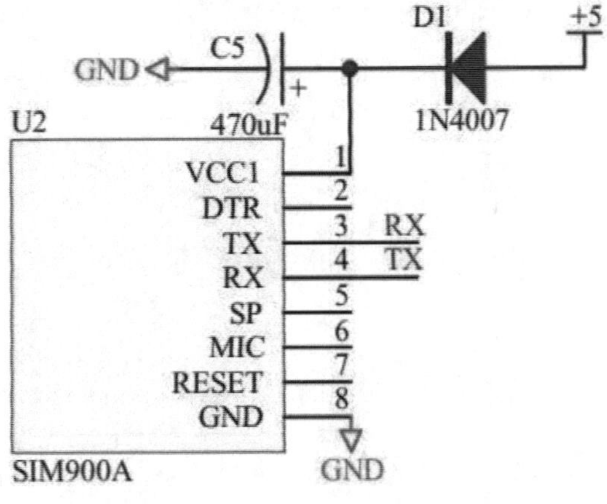

Figure 9. GSM module circuit

IOP Conf. Series: Materials Science and Engineering **630** (2019) 012010 doi:10.1088/1757-899X/630/1/012010

5. Physical drawing

As shown in the figure is the physical diagram of the intelligent alarm system, the annotations in the diagram correspond to each component one by one.

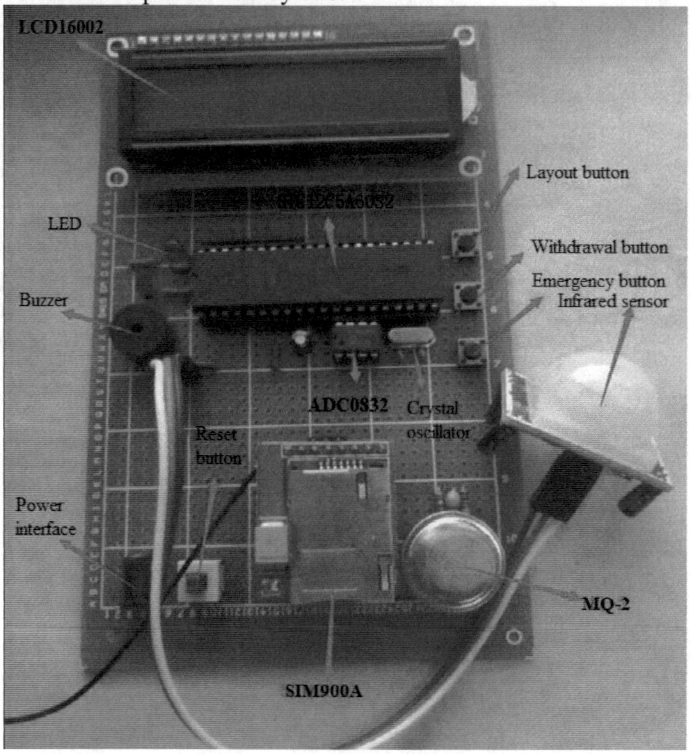

Figure 10. Physical drawing

6. Program design

This design mainly uses the modular programming method, divides into the main module, the interrupt module, the initialization module, the short message module, the delay module and so on.

C language is used to write the program. According to the design flow chart, after the initialization of the power supply, each module enters the working state. after the smoke sensor, the infrared sensor monitors the signal or the button to monitor and judge, the buzzer carries on the alarm, the GSM carries on the short message notification.

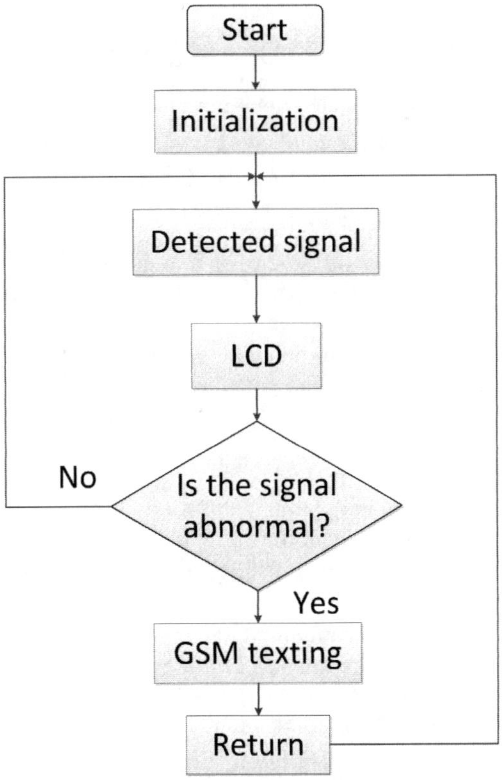

Figure 11. Block diagram

7. Experiment

7.1 Experimental purpose
Through the test of different performance single-chip microcomputers, the optimal single-chip microcomputer that meets the requirements can be selected.

7.2 Experimental protocol
Under the condition that other external conditions are guaranteed to be unchanged, the STC89C51 single-chip microcomputer and the STC12C5A60S2 single-chip microcomputer are selected for comparison test.

Based on the circuit board on which the STC12C5A60S2 is installed, the STC12C5A60S2 single-chip microcomputer is replaced with the 89c51 single-chip microcomputer to test the running speed.

7.3 Experimental results
Through experimental verification, after replacing the MCU, it can be clearly compared that the speed of running STC12C5A60S2 is 8 to 12 times that of STC89C51.

8. Conclusion
According to the test results, using a better performance single-chip microcomputer can improve the operating speed of the system without affecting normal use, and the safety performance and reliability meet the requirements.

The price of the parts in this design is relatively low, and there is no pollution during the work, which is in line with the requirements of sustainable development.

IOP Conf. Series: Materials Science and Engineering **630** (2019) 012010 doi:10.1088/1757-899X/630/1/012010

Acknowledgments

This work was supported in part by Key R & D project of Shandong Province under Grant 2017CXGC0215, Grant 2017CXGC0810 , Grant 2018CXGC0908 and in part by Shandong Province Natural Science Foundation under Grant ZR2017LEE010, and in part by Research and Development Innovation Plan Project of Agricultural Machinery equipment in Shandong Province under Grant 2017YF047

Thank you, Mr. Zhang Chengliang, for your careful guidance, and thank you, colleagues and friends, for your encouragement and support. Along the way, from your body I harvest countless, but no return, here to express my deep gratitude to the teachers! Thank your parents! Thank you to all the students who have helped and supported me, my friends! You're all my solid pillars and backers. Thank you for applauding me when I succeed and comfort me when I fail. I will never forget your concern, trust, encouragement and help in your daily study, work and life!

Reference

[1] Li H 2018 Discussion on the Application advantage and Development of single Chip Microcomputer in Electronic Technology *J. Sci. Technol. Innovation Guide* **15** 95-6

[2] Zhao X H, Hao Y,Liu Y and Lv X P 2013 Design of Infrared alarm system *J. Applied Sci. Technol.* **40** 47-52

[3] Sun B,liu S C,Guo S,Li H S and Zhang Z Y 2018 Design of smoke alarm device based on AT89C51 single Chip Microcomputer *J. Laboratory Sci.* **21** 45-50

[4] Sheng J and Zhang L L 2015 Design of Home Wireless Fire alarm system based on single Chip Microcomputer *J. Fire Sci. Technol.* **35** 1421-4

[5] Zhang Q Q and Zhao Q N 2015 Design of Fire alarm system based on MQ- 2 Sensor *J. Value Eng.* **34** 96-8

[6] Yi J Q,Huang Y, Liao H H,Tan J J and Jiang S L 2015 Pyroelectric Infrared Sensor and its Application in personnel Counting system *J. Infrared Laser Eng.* **44** 1186-92

[7] Pan X W,Wang Z R and Chen M W 2016 Infrared burglar alarm system based on single Chip Microcomputer and GSM Module *J. Journal of Hunan Institute of Sci. Technol.* **37** 26-8

[8] Dong W H,Wang Li, Yu G Z and Mei Z B. Design of Wireless Automatic Fire Alarm System. *J. Fire Sci. Technol.* **35** 552-5

IOP Conf. Series: Materials Science and Engineering **630** (2019) 012011 doi:10.1088/1757-899X/630/1/012011

Cognitive model of digital production development

Alla Brom and Yuliya Maslennikova

Bauman Moscow State Technical University, Moscow, Russian Federation, 105005

maslennikova.yuliya@yandex.ru

Abstract. The aim of the work is to develop a cognitive model that reflects clear, structured, formalized, obvious relationships between the social – economic factors of digital production. The main problems of industrial enterprises are reflected, digital industrial technologies are studied, the main factors of influence on their development are considered, methods of assessing the impact of factors on the development of technologies are studied, and the cognitive model is developed. The practical significance of the paper lies in the possibility of using the presented cognitive model by industrial enterprises. This will improve the efficiency of digital technology implementation and the use of available resources.

1. Introduction

The global industry is entering an era of industrial change. Nowadays, we increasingly associate production systems with such concepts as "Smart production", "Internet of things", "Big Data", etc. [1-2]. Industry 4.0. entails the emergence of new industries and changes in existing ones, enterprises become more flexible and adaptive. Many authors raise the topic of industrial transformation within the framework of digitalization of economy in their works. [3-4].

The emergence and spread of digital solutions in the Russian Federation are at a low level. The reasons may be in the complexity of implementation, the historical technological gap from Western partners, a low level of investment in digital innovation, digital inequality inside the country, reducing of healthy competition. To increase competitiveness in the global market it is necessary to move Russian production systems to a new level. The sources of the evolution of production systems are not only automation technologies, but also social changes generated by the transformation of the digital economy. [5-6]. Moreover, these processes are interrelated. However, these factors are weakly obvious and not formalized. For clarity, we consider the mentioned problem by the example of industrial company OOO «Voenno-promishlennaya companiya» (OOO «VPC»). The company develops and manufactures armoured vehicles, multi-purpose and lightly armoured vehicles, special vehicles for security and law enforcement.

An original and innovative aim of this paper is to model the socio-economic processes of digital production. Industrial enterprises will be able to discern imperfections and gaps in the strategy of digital technologies implementing and to improve the transition to a new production level efficiency.

2. Research Method

Many authors use statistical-mathematical methods to evaluate the effect of various factors. Deterministic and stochastic methods are simple methods of calculating the factors influence. To calculate factors that do not have a direct impact on the resulting indicator regression and linear programming, correlation and ranking methods are used. [7]. However, these methods do not provide

a full picture of interactions between the factors themselves and the development of the resulting indicator.
The authors will use the method of system dynamics, namely the construction of a cognitive model. American scientist R. Axelrod proposed the methodology of cognitive modelling for the analysis and decision-making in difficult certain situations. [8]. The essence of the cognitive approach is reflected in the preparation of a simplified model that shows the complex and unpredictable trends of the situation development. To analyse the behaviour of a complex system, a scheme of cause-and-effect relationships is constructed, which is called a cognitive map or a directed graph. Connections can be positive and negative, what means that the increase (decrease) of one factor leads to an increase (decrease) of the other, and the increase (decrease) of one factor leads to a decrease (increase) of the other. Firstly, to build relationships, the target factors of influence on the resulting indicator are selected, then additional factors and leverages. After that, it is necessary to find out what factors and leverages have a connection. The completion of the construction is a functional graph; here we characterize the relationships by quantitative or linguistic variables that take values in the interval [0; 1]. Further, the cognitive method allows us to analyse the system from two sides: in terms of its structure and in terms of its dynamics. Static analysis is studying the structure of mutual influences of cognitive map factors. This analysis allows us to identify the factors with the strongest impact on the result indicator, i.e. the factors whose values need to be changed. Dynamic analysis shows the development of the situation in time.

3. Results and discussions

3.1. Machine-to-machine technology

The most striking trend among the digital production technologies is the industrial Internet of things, which includes the technology of machine interaction (Machine-to-machine). [9].
Machine-to-machine (M2M) is a technology that allows enterprise devices to communicate with each other without human assistance, using wired and wireless sensors. The technology M2M realize the opportunity to improve the quality of products and reduce the degree of defects, in equipment diagnostics and optimization of repair time, eliminating unnecessary preventive maintenance, to avoid downtime, to ensure effective service, including the customer's site and to reduce operating costs of production. The technology allows transferring large data in real time. It helps to monitor constantly the quality of ongoing assembling processes, to simplify and speed up logistics processes, to minimize unwanted costs.
OOO «VPC» faces the same problems, as well as other industrial enterprises, namely: low level of competitiveness, unjustified size of expenses, obsolescence of fixed assets, lack of innovative activity, big expenses of a time resource, low level of service. The digitalization of production can help plants to overcome these barriers. It is interesting to investigate the influence of social-economic factors on the introduction of M2M technology on the example of wheeled combat vehicles assembling.
Assembling as a component of the production process is a complex multi-level process described by an assembling scheme. Assembling of wheeled combat vehicles is classified into nodal, aggregate and general. Thanks to M2M technology, a model of the assembling process will be transferred to all necessary equipment. The technological process model of assembling is a complex of works description of installation and connection of components in order to obtain a design that meets the specified requirements. It is a computer model that is implemented in the information environment. During the assembling process, it describes the changes in assembling unit, the interaction of operations, the transitions from one type of work to another. The technological process of assembling design scheme is described by the following formula [10]:

$$S(A1T)p1TA\ Sn(T1A) \rightarrow Sp(T1A) \quad (1)$$

Where S(A1T) is the technological model of an assembling unit, the initial product description;

Sn(T1A) is a process design model formed by the elements of the generating environment technological design;
Sp(T1A) is the model of technological process - the result of transformation implementation;
p1TA is a procedural-algorithmic environment of technological design.
Such a scheme describes the process of obtaining the technological process model based on the model of the assembling unit. It will be transmitted to each equipment by using sensors.

3.2. Cognitive model composition
A development of M2M technology at OOO «VPC» will be called as the level of digitalization of production in the cognitive model. We select the target factors affecting the level of digitalization of production: Implementation period; Level of investment; Level of personnel qualification.
Authors start building the cognitive model with separate feedback loops. Figure 1 shows the first cycle reflecting the impact of the level of investment.

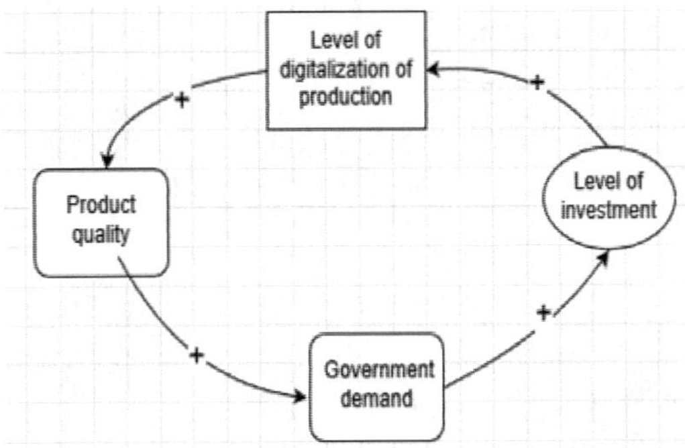

Figure 1. The impact of the level of investment on the development of digitalization of production.

It is an amplifying loop, which reads as follows: "The higher the level of investment, the higher the level of digitalization of production; the higher the level of digitalization of production, the higher the quality of products, the higher the quality of products, the higher the government demand, which causes an increase of investment. Let us continue the construction of the contour with the target factor «Level of personnel qualification» in Figure 2. In addition to a cycle, this graph shows the leverages «Age», «Availability of additional training» and «Quality of education». Furthermore, there is a negative connection, so the loop is stabilizing.

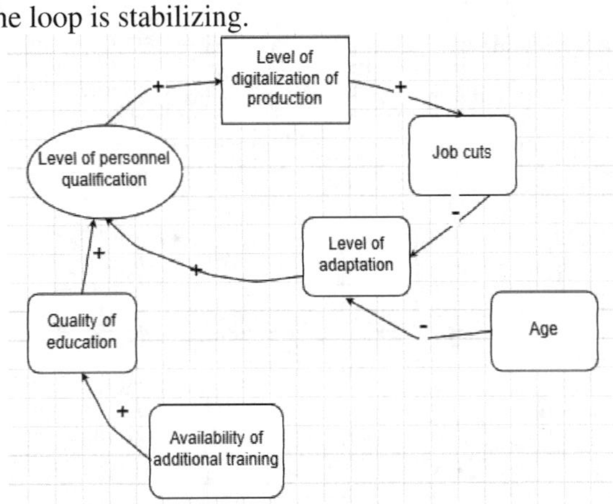

Figure 2. The impact of Level of personnel qualification on the development of digitalization of production.

We need to combine these contours taking into account the following target factor – implementation period. It is influenced by previous factors – the level of personnel qualification and the level of

IOP Conf. Series: Materials Science and Engineering **630** (2019) 012011 doi:10.1088/1757-899X/630/1/012011

investment. When the contours are combined, we can see an additional link between the level of investment and the availability of additional training. With adding a new factor and a new relationship, instead of two loops we have five cause-and-effect contours. Next, we add letter designations to factors and quantitative weights to the identified relationships. The result is a cognitive model. (Figure 3.)

The following step, we make up a matrix of effects. For example, the relationship between "Level of digitalization of production " (A) and "Product quality0" (E) will be reflected in the matrix as S(A;E) = +0.7, where S is a general designation of the whole matrix, A and E are the designation of factors by row and column, respectively. Or, the relationship between "Level of personnel qualification" and "Implementation period" will be reflected as S(D;B) = -0,5. All values are shown in table 1.

What happens if we change one of the factors by a certain amount? Figure 4 shows the development of each factor in dynamics with an increase of one of the leverages «The government demand» by 10%. The graph clearly shows the qualitative relationships between various factors and the level of digital technologies development at the enterprise.

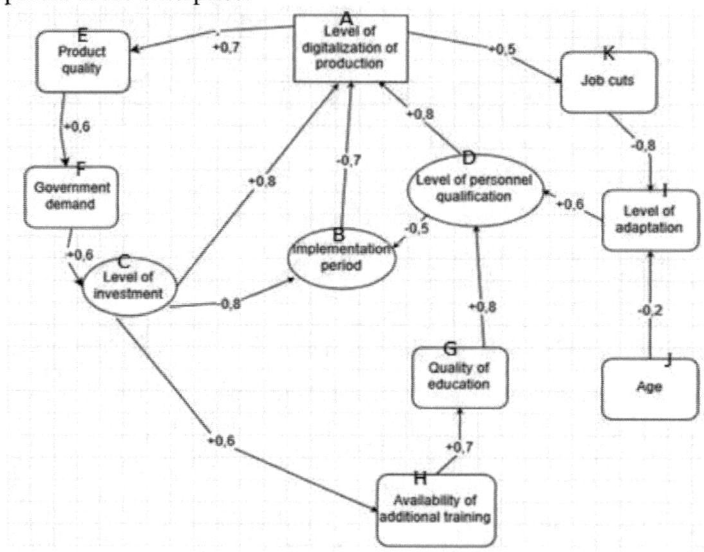

Figure 3. Cognitive model of digital production development.

Table 1. Matrix of factors interaction.

	A	B	C	D	E	F	G	H	I	J	K
A	0	-0,7	0,8	0,8	0	0	0	0	0	0	0
B	0	0	0	-0,5	0	0	0	0	0	0	0
C	0	0	0	0	0	0,6	0	0	0	0	0
D	0	0	0	0	0	0	0,8	0	0,6	0	0
E	0,7	0	0	0	0	0	0	0	0	0	0
F	0	0	0	0	0,6	0	0	0	0	0	0
G	0	0	0	0	0	0	0	0,7	0	0	0
H	0	0	0,6	0	0	0	0	0	0	0	0
I	0	0	0	0	0	0	0	0	0	-0,2	-0,8
J	0	0	0	0	0	0	0	0	0	0	0
K	0,5	0	0	0	0	0	0	0	0	0	0

Indeed, with a Government demand increase by 10%, the flow of investment increases by 6% and there is an increase of the level of digitalization of production by 4.8 %. The product quality increases, there is a possibility of additional training. However, let us pay attention to the 10th period. It is here that the development of digitalization of production stops. It turns out that despite the growth of government demand, increasing funding in technologies, the gradual reduction of implementation time and improving the quality of education, the level of personnel qualification, as the most important social factor is not growing. This is so due to the continuously decreasing level of adaptation, which is influenced by the age of the workers and the factor of job cuts. In addition, implementation period is almost not accelerated, because it is also depends on the staff. This leads to the conclusion that managers are building digital policy incompetently and do not care about the digital culture within the company. Only people, namely highly qualified personnel is able to make the enterprise digital. Therefore, the company should consider the introduction of technology as a system that depends on many factors and does not start working immediately. The company needs to pay close attention to the components of social processes, because a person is the most important resource of the enterprise.

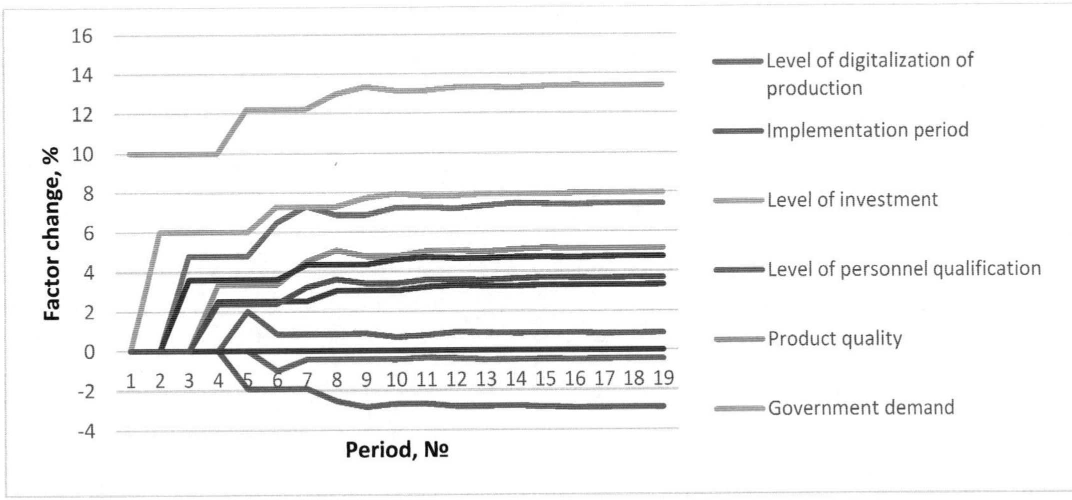

Figure 4. Graph of dependence of the social-economic factors of the digital production.

4. Conclusion

A cognitive model of implementation of the digital technology M2M to the industrial enterprise is developed. It has a clear structure and qualitative relationship between factors influencing the evolution of digital production technologies. The practical application of this model will improve the efficiency of implementation of digital industry.

References

[1] I. Alpackaya, D. Alpackiy, *Perspectives and consequences of implementation and development digital economy,* MATEC Web of Conferences **193**, 05087, (2018).

[2] E. Lyapuntsova, Y. Belozerova, , I. Drozdova, G. Afanas'Ev, E. Okunkova, *Entrepreneurial Risks in the Realities of the Digital Economy,* MATEC Web of Conferences, **251**,06032, (2018)

[3] M. James. *Changing roles in the electronic age - the library perspective*, Libr. Acquis. Pract. and Theory. № 1. P. 15-21. (1998).

[4] V. Akberdina, A. Kalinina, A. Vlasov, *Transformation stages of the Russian industrial complex in the context of economy digitization* Problems and Perspectives in Management 16(4), p. 201-211, (2018).

[5] E.A. Tishina, , E.Y. Rezantseva, , D.V. Reut, *The concept of digital transformation of the society,* Proceedings of 2017 10th International Conference Management of Large-Scale System Development, MLSD, 8109697, (2017).

[6] Y.L. Maslennikova. *Problems of low retraining of specialists in the transition to the sixth*

technological mode. Management systems for the full life cycle of high-tech products in mechanical engineering: new sources of growth: materials of all-Russian science.-prakt. conf. M.: PH BMSTU, P. 116-119. (2018).

[7] V. I. Loyko, E. V. Lutsenko, A. I. Orlov, *Modern digital economy.* Krasnodar: Kubgau, **508**, (2018).

[8] R. Axelrod, The Structure of Decision: Cognitive Maps of Political Elites. Princeton. University Press, (1976).

[9] A.V. Manzhirov, *Some problems in mechanics of growing solids with applications to AM technologies.* Journal of Physics: Conference Series 991(1),012056, (2018).

[10] O. S. Samsonov, *Modelling of processes of engineering and design to the Assembly of aircrafts.* Technology of mechanical engineering.No. 9. P. 18-26. (2007)

IOP Conf. Series: Materials Science and Engineering **630** (2019) 012012 doi:10.1088/1757-899X/630/1/012012

The basic set of test problems for ODE system solvers

Vladimir Manichev, Dmitriy Zhuk and Emiliy Feldman

BMSTU, Robotics and Complex Automation Department, Computer Aided Design and Engineering, 105005 Moscow, Russia

feldlime@yandex.ru

Abstract. Known mathematical packages, MATLAB, Maple, Mathematica, MathCAD and others may get wrong, often plausible, the result of numerical solution of ODE systems with low, given by default, the requirements for mathematical accuracy of the results of the numerical solution of ODE systems. Since ODE system parameters obtained usually experimental, with a low mathematical precision, so the requirements for the precision of the results of mathematical solutions of ODE systems is low (for example, in MATLAB package required precision is 0.001). The article offers a basic set of tests to assess the range of applicability of the relevant solvers. The basic set of test problems for ODE solvers systems include linear ODE systems with the known analytic solution and nonlinear systems with known graphics solution. Presented comparative results of solutions for proposed problems using MATLAB solvers and *manzhuk* program from a library of standard mathematical programs SADEL (Sets of Algebraic and Differential Equations solvers Library), which has been designed for reliable and accurate solving of systems of linear algebraic equations (LAE) and ODE systems. The results can be used in mathematical modeling of dynamic systems described by ODE systems.

1. Introduction

The programs for solving ODE systems use three classical problem statements for solving ODE systems:

1. The Cauchy normal form in the space of differential state variables (explicit form) [1]:

$$d\boldsymbol{X}/dt = \boldsymbol{F}(\boldsymbol{X}, t)$$

Where $\boldsymbol{X}$ is the vector of the coordinate basis of differential variables of the state of dimension m; $\boldsymbol{F}$ is the vector function of the right parts of the dimension m; t is the independent variable (usually time). Initial conditions specified as $\boldsymbol{X_0} = \boldsymbol{X}(T_0)$, time of integration is $t = [T_0, T_E]$, T_0 is the time of the beginning the integration, T_E – time of completion of integration.

2. Differential-algebraic form of different indices in the full coordinate basis (semi-explicit form) [2, 3]:

$$\begin{cases} d\boldsymbol{X}/dt = \boldsymbol{F}(\boldsymbol{X}, \boldsymbol{Y}, t) \\ \boldsymbol{G}(\boldsymbol{X}, \boldsymbol{Y}) = \boldsymbol{0} \end{cases} \quad (1)$$

Where $\mathbf{Y}$ – vector of algebraic variables of the full coordinate basis of differential-algebraic variables of dimension k, $\boldsymbol{G}$ is the vector function of dimension k. The agreed initial conditions are now specified as $\boldsymbol{X_0} = \boldsymbol{X}(T_0)$, $\boldsymbol{Y_0} = \boldsymbol{Y}(T_0)$. This statement (1) of the task of ODE systems solutions are

IOP Conf. Series: Materials Science and Engineering **630** (2019) 012012 doi:10.1088/1757-899X/630/1/012012

widely used in the mathematical and numerical modeling automatic regulation and control systems, for example, MATLAB-SIMULINK [4], MVTU [5, 6] and others.
3. The differential-algebraic form in the full coordinate basis, i.e. system of differential-algebraic equations (DAE) (implicit form):

$$\boldsymbol{G}(\boldsymbol{X}, d\boldsymbol{X}/dt, \boldsymbol{Y}, t) = \boldsymbol{0} \tag{2}$$

Where $\boldsymbol{G}$ – vector function of dimension $(m + k)$. The initial conditions are the same as in the previous case. Statement (2) is used in mathematical and numerical modeling of electronic and mechanical engineering systems, for example, the ODE systems solvers SPICE and ADAMS. The task (2) in the space of differential algebraic variables was set in the full coordinate basis by L. Petzold in 1982 and she developed the DASS (Differential Algebraic Systems Solver) program based on the Gear method with replacement of derivatives according to the Gear formula [7]:

$$d\boldsymbol{X}/dt = BDF(\boldsymbol{X}, h)$$

Where BDF – Backward Differential Formula, h – integration step. The resulting system of nonlinear algebraic equations was solved by the Newton method at each integration step.
The task of solving ODE systems in the extended space of differential-algebraic variables without transformations of the initial ODE systems is based on the statement (2) of the problem and has the form:

$$\boldsymbol{G}(\boldsymbol{X}, \boldsymbol{PX}, \boldsymbol{Y}, t) = \boldsymbol{0} \tag{3}$$

Here $\boldsymbol{PX} = d\boldsymbol{X}/dt$ vector of derivatives of differential state variables with respect to time of dimension m.
Considering the fact that some matrix-vector transformations on a computer are not equivalent due to the limited bit grid of computers, the classical forms of representation of ODE systems should be used mainly for theoretical studies, provided that all transformations of the original DAE systems are performed analytically or in symbolic form.
The stability domain of any explicit method is limited, therefore, implicit methods are recommended for solving hard problems [1, 2]. But implicit methods also have their drawbacks. Using the implicit method for solving stiff and locally unstable problems may lead to qualitatively wrong solution [8]. It is undesirable to use implicit methods for modeling hybrid dynamic systems containing discrete and relay elements, since in this case the Jacobi matrix must be updated frequently. Finally, implementing implicit methods is much more complicated than explicit ones. Therefore, attempts to build explicit methods that are effective for hard problems do not stop [8]. This article discusses the basic set of test ODE systems, which includes both stiff and nonstiff ODE systems and comparing solvers of MATLAB with *manzhuk* program from the library of standard mathematical programs SADEL in this set of tests.

2. Methods for solving ODE-DAE systems

There are implemented 8 methods for solving ODE-DAE systems in the MATLAB. The choice of methods for solving ODE-DAE systems in the SADEL library is explained below.
Since the explicit methods for solving systems of ODE and DAE are not universal and do not meet the requirements of the stability in the solution with variable step integration of the stiff ODE systems, then as basic integration methods for SADEL library were selected implicit methods of integration [2]. Implicit methods for solving ODE systems are reduced to iteratively solving nonlinear algebraic systems. In the software implementation of these methods, two main types of errors in the numerical solution of ODE systems should be distinguished – global error (or qualitative error) and local error (or quantitative error).
In theoretical terms, the properties of ODE-DAE systems and methods for their solution are considered in relation to linear inhomogeneous ODE systems of the form:

$$d\boldsymbol{X}/dt = A * \boldsymbol{X} + \boldsymbol{H}(t) \tag{4}$$

Where A is a constant real matrix with size $m \times m$, $\boldsymbol{H}(t)$ is a known vector function of time.
If the integration method is not suitable for solving the system (4), it is unsuitable for system (3) too. Denote the determinant of the matrix A through $det(A)$. The spectrum of eigenvalues is the set of all eigenvalues λ_i of matrix A, $i = 1,2,\dots,m$. In the general case $\lambda_i = \mathrm{Re}(\lambda_i) + j * Im(\lambda_i)$ is a complex value. It is known that if the eigenvalues are different, then the general solution of system (4) will be a linear combination of fundamental solutions $c * e^{Re(\lambda_i)t} * cos(Im(\lambda_i)t)$ and $c * e^{Re(\lambda_i)t} * sin(Im(\lambda_i)t)$, c is some constant. The analytical solution of system (4) can be represented as the sum of the individual fundamental solutions of a homogeneous system of ordinary differential equations with a matrix A and the private solution of the heterogeneous system (4):

$$\boldsymbol{X}(t) = \sum_{i=1}^{m} c_i e^{\lambda_i t} \boldsymbol{X}_i + \boldsymbol{Z}(t)$$

Where c_i – constant values; $\boldsymbol{X}_i$ – own vector of matrix A corresponding to the eigenvalue λ_i; $\boldsymbol{Z}(t)$ – private solution of the system (4). For these solutions five areas of the complex plane λ_i reflecting the five classes of local problems of integration of differential equations (Figure 1) can be distinguished.
In general, in a wide variety of ODE- DAE systems can be isolated five base classes of tasks for basic solutions of linear homogeneous ODE shown in Figure 1 and requiring the use of various methods for solving.

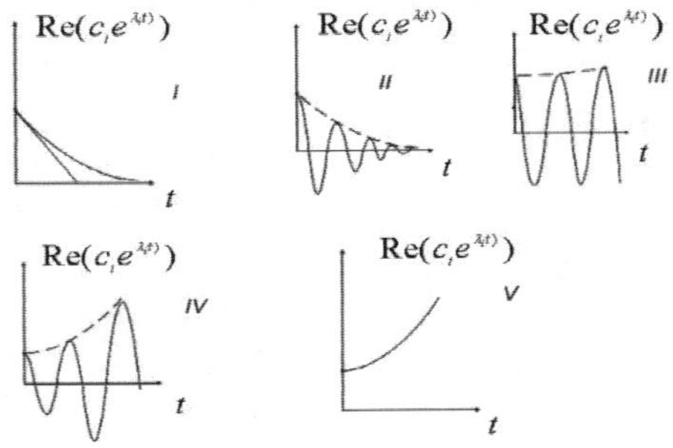

Figure 1. Five base classes for fundamental solutions of linear homogeneous systems of ODE.

The classical control of the local integration error will ensure the accuracy of the solution, and to control the global integration error it is necessary to ensure the appropriate stability of the integration methods. To ensure obtaining a qualitatively correct solution for the above classes of problems, the integration method must be stable for stable ODE systems (I and II classes of tasks) and should be unstable for unstable ODE systems (IV and V classes of problems), therefore it is necessary that the integration methods be appropriately AL-stable (AL-stability – Absolute stability in Left half-plane of the complex stability plane of numerical integration), i.e. absolutely stable strictly in the left half-plane of the complex plane of stability of the integration methods [9] .
With low requirements to the precision of solving ODE systems, this area of absolute stability guarantees the preservation of a qualitatively correct solution with increasing integration step in the plane $h\lambda$, since with low integration accuracy the integration step can become so large that non-AL-stable integration methods can obtain incorrect solution trajectories. Implicit one-step Runge-Kutta integration methods meet the AL-stability requirement [10]. Algorithms for the implementation of implicit one-step integration methods of Runge-Kutta type are based on the joint solution of systems of nonlinear algebraic equations $\boldsymbol{F}(\boldsymbol{X},\boldsymbol{Y},\boldsymbol{PX},t) = \boldsymbol{0}$ for solved DAE system of the form (3) and systems of linear algebraic equations for integration methods $\boldsymbol{H}(\boldsymbol{X},\boldsymbol{PX},h_i) = \boldsymbol{0}$, formed at the corresponding stages of the integration methods, relative to differential-algebraic variables $\boldsymbol{X},\boldsymbol{Y}$ and derivatives of differential variables $\boldsymbol{PX}$ [11]. Were developed DAbc formulas (as the development of Ab^{T}c Butcher methods [12]) for S-phasic implicit methods for solving DAE systems in an extended coordinate basis:

$$\boldsymbol{H}_{ii}(\boldsymbol{PX}_{ii},\boldsymbol{X}_{ii},\boldsymbol{X}_{i-1},\boldsymbol{PX}_{i-1},h_i) = h_i \sum_{j=1}^{S} d_{iij}\boldsymbol{PX}_j - \sum_{j=1}^{S} a_{iij}\boldsymbol{X}_j - b_{ii}\boldsymbol{X}_{i-1} - h_i c_{ii}\boldsymbol{PX}_{i-1} = \boldsymbol{0}$$

$$\boldsymbol{X}_i = \boldsymbol{X}_s, \boldsymbol{PX}_i = \boldsymbol{PX}_s, ii = 1, \dots, s, t_i = t_s$$

Where s – the number of stages of the corresponding one-step method ; h_i – i-th integration step; d_{iij}, a_{iij}, b_{ii}, c_{ii} – method parameters. Based on these formulas, three integration methods were developed and implemented:

- M1 – A-stable one-step implicit method of the first order of accuracy;
- M2 – AL-stable one-step implicit method of the second order of accuracy;
- M3 – AL-stable two-stage implicit method of the fourth order of accuracy.

The main drawback of AL-stable methods is that they do not satisfy the requirements of L – stability and sometimes give "false fluctuations" in the solution, so these methods are not widely used in practice [2]. The solution to this problem is given in [10].

3. Basic set of test ODE systems

3.1. Test problems with a well-known analytical solution

It is proposed to develop 5 test linear homogeneous ODE systems 1) – 5) with analytical solutions corresponding to the classes of tasks shown in Figure 1 as the basic set of test problems. Given the test linear system of ODE of the 2nd order:

$$d\boldsymbol{X}/dt = A * \boldsymbol{X} + \boldsymbol{B}$$

Where $\boldsymbol{X}$ is the vector of unknowns of system , A is the matrix of constant coefficients of a system of size 2×2, $\boldsymbol{B}$ is a constant vector.

3.2. Test problems with a graphical solution:

6) An example of the calculation of a tough second-order ODE system (MU is a stiffness parameter) – Van der Pol's test. Correct solution for this test shown in Figure 2.

$$\begin{cases} \dfrac{dx_1}{dt} = x_2 \\ \dfrac{dx_2}{dt} = -x_1 + MU * (1 - {x_1}^2) * x_2 \\ x_1(0) = -1, x_2(0) = 1 \\ t \in [0, 4.2MU] \end{cases}$$

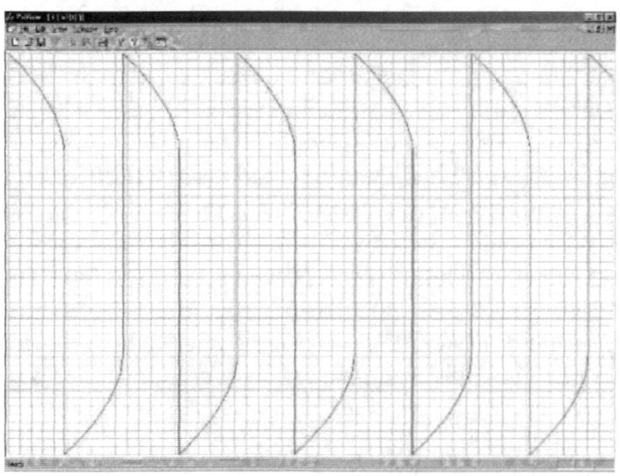

Figure 2. Reliable solution results for test 6.

IOP Conf. Series: Materials Science and Engineering **630** (2019) 012012 doi:10.1088/1757-899X/630/1/012012

7) Example of electronic circuit simulation with a multi-period solution (5th order ODE system).

8) 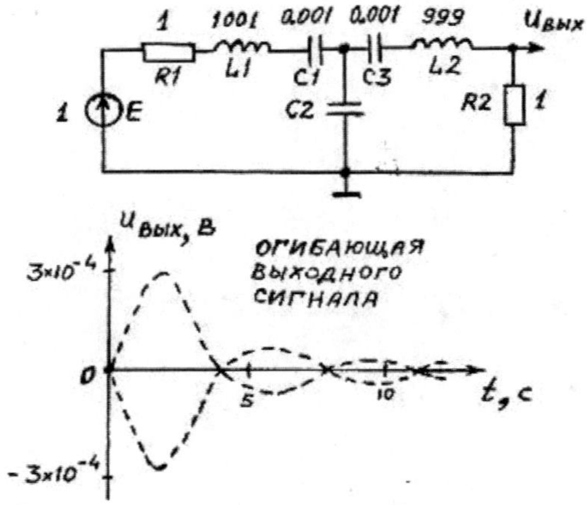

Figure 3. Electronic circuit for test 7

The ODE system for this test is:

$$\begin{cases} kr = \dfrac{ku}{ki}, kc = kt * \dfrac{ki}{ku}, kl = kt * \dfrac{ku}{ki} \\ \dfrac{dx_1}{dt} = \dfrac{x_4}{0.001} * kc \\ \dfrac{dx_2}{dt} = \dfrac{x_5}{0.001} * kc \\ \dfrac{dx_3}{dt} = \dfrac{x_4 - x_5}{kc} \\ \dfrac{dx_4}{dt} = \dfrac{ku - x_1 - x_3 - kr * x_4}{1001} * kl \\ \dfrac{dx_5}{dt} = \dfrac{-x_2 + x_3 - kr * x_5}{999} * kl \\ x_1(0) = 0,\ x_2(0) = 0,\ \ x_3(0) = 0 \\ x_4(0) = 0, x_5(0) = 0 \\ t \in [0, 12560 * kt] \end{cases}$$

The output voltage corresponds to $x_5(t)$.

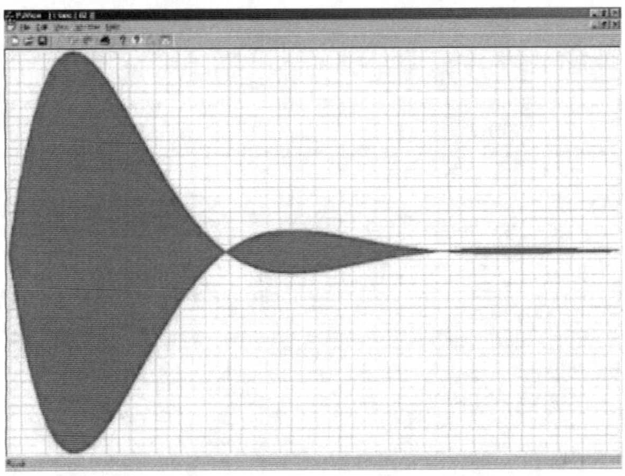

Figure 4. Correct solution for test 7.

9) Nonlinear stiff ODE system for mathematical modeling of real laser processes.

$$\begin{cases} \frac{dx_1}{dt} = -x_1 * (\alpha * x_2 + \beta) + \gamma \\ \frac{dx_2}{dt} = x_2 * (p * x_1 - \sigma) + \tau * (1 + x_1) \\ x_1(0) = -1, x_2(0) = 0 \\ t \in [0, 10^6] \\ \alpha = 1.5 * 10^{-18}, \beta = 2.5 * 10^{-6}, \gamma = 2.1 * 10^{-6}, \\ p = 0.6, \sigma = 0.18, \tau = 0.016 \end{cases}$$

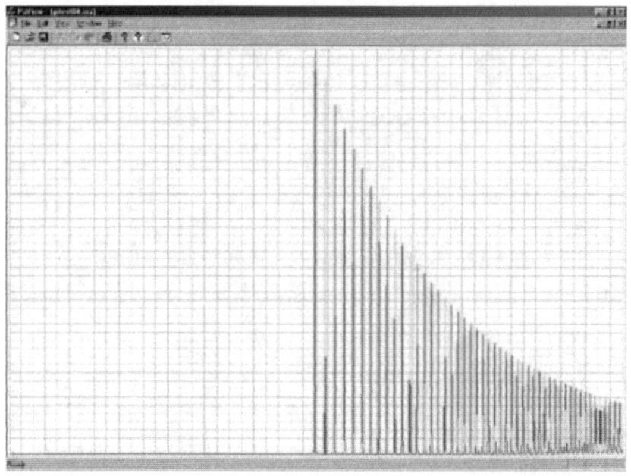

Figure 5. The correct solution for $x_2(t)$ in test 8.

Waivers for problems with graphical solution are time of calculation more than 10 seconds or the wrong plot of the solution.

4. Calculation results

4.1. Test problems with a well-known analytical solution :

Failures for tasks with an analytical solution were recorded if the maximum relative error was more than 0.001 in a given integration interval. These tasks were used to test manzhuk solver and different MATLAB solvers.

The final test results are shown in Table 1.

Table 1. The final results of the tests with known analytical solution.

Program for solving ODE systems	Tests				
	1)	2)	3)	4)	5)
manzhuk (M3)	+	+	+	+	+
MATLAB ode45	+	+	-	-	-
MATLAB ode23	+	+	-	-	-
MATLAB ode113	+	+	-	-	-
MATLAB ode15s	+	+	-	-	-
MATLAB ode23s	+	+	-	-	-
MATLAB ode23t	+	+	-	-	-
MATLAB ode23tb	+	+	-	-	-
MATLAB ode15i	+	+	-	+	+

Sign (+) means a correct solution of the test with accuracy $eps = 0.001$.
Sign (-) means failure with accuracy $eps = 0.001$.
It should be noted that the qualitative solution for all MATLAB methods was correct. Maximum errors were obtained on test 3.

4.2. Test problems with a graphical solution
In view of the results published in [10, 13, 14], Table 2 shows the summary comparison solvers of stiff ODE systems of the aforementioned SADEL library, from well-known libraries and packages of mathematical programs.

Table 2. The results of the tests with known graphical solution.

Program for solving ODE systems	**Tests**				
	6)	6)	7)	7)	8)
	MU= 1e6	MU=1e9	Kt=1 Ki=1 Ku=0.01	Kt=1e-103 Ki=1, Ku=1	The real laser parameters
manzhuk (M3)	+	+	+	+	-++
NAG C-Library	-	-	-	-	+
IMSL C-Library	-	-	-	-	-
Mathcad	+	-	-++	-	+
MATLAB ode15s	+	-	-++	-	+
Maple	-	-	+	+	-++
Mathematica	+	-	+	-	+
SPICE-Multisim	-	-	-++	-	-+
OrCAD- PSPICE	-	-	-+	-	-
SPICE-SYMICA	-	ND	-+	-	ND
OpenModelica	-	-	-	-	-

The sign (+) means the correct solution of the test using the method recommended for solving stiff ODE systems, with an accuracy $eps = 0.001$.
The sign (-) indicates an incorrect test solution using the method recommended for solving hard ODE systems with default parameters without warning the user and refusing to get the right solution.
The sign (-+) means an incorrect test solution by the method recommended for solving hard ODE systems, with default parameters without warning the user and obtaining the correct solution by setting appropriate methods and integration parameters and increasing the calculation time.
Sign (-++) means an incorrect test solution by the method recommended for solving hard ODE systems, with default parameters without warning the user and obtaining the correct solution by setting only the parameter specified accuracy of the method of integration and increasing the calculation time.
ND – no data.
Figure 6 shows as an example the result of a qualitatively incorrect solution for the 1st variable of test 7) by the implicit method ode15s and correct solution by method M3 of program *manzhuk*. There were no warnings about the incorrect decision in the ode15s solver.

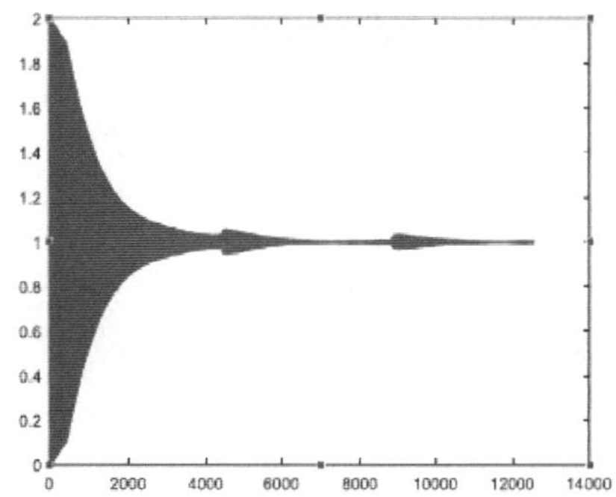

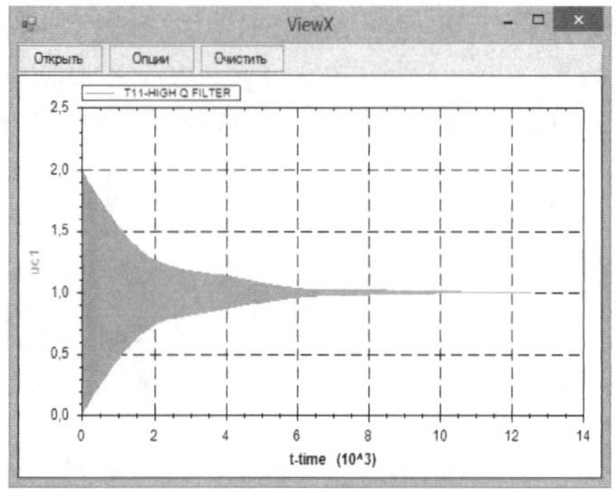

Figure 6. Solution for the 1st variable of test 7): by ode15s; by method M3 of program *manzhuk* (right).

5. Conclusion

Most of the problems of solving systems of ordinary differential equations arising in practice cannot be solved analytically [15]. Numerical methods descend to the seventeenth century – to Newton, who proposed using them to calculate the trajectories of comets. Manual calculations continued to be used until the 60s of the twentieth century, before the appearance of the first solver programs written in FORTRAN for the first computers [16]. Since that time, development of test problems for solving ODE systems has been carried out. Classical sets of test considered tests from books [1, 2, 17]. The first test comparison of the MATLAB ODE solvers is given in [18].

In MATLAB we can always choose the method and adjust its parameters to obtain qualitatively correct results, but most users are not experts in numerical methods, so the comparison should be made for the recommended default parameters of integration ODE solvers.

For specific classes of problems to be solved, it is necessary to have sets of test practical and mathematical problems with a known deliberately reliable and accurate solution for the correct selection and adjustment of the corresponding solvers of ODE systems and the correct evaluation of the obtained results of numerical simulation.

This article presents the results obtained in the continuation of the above works. Further studies focused on developing effective methods and algorithms for the automatic integration step selection for AL-stable implicit methods for the numerical solution of high- and ultra-high-dimensional ODE-DAE systems and testing these new solvers. The relevance and importance of these works is indicated in the [19, 20].

References

[1] E. Hairer, S.P. Norsett, G. Wanner, *Solving ordinary differential equations I. Nonstiff Problems.* Springer, **528** (2008)

[2] E. Hairer, G. Wanner, *Solving ordinary differential equations II. Stiff and Differential-Algebraic Problems.* Springer, **614** (2010)

[3] A.D. Polyanin, I.K. Shingareva, *Non-linear blow-up problems for systems of ODEs and PDEs: Non-local transformations, numerical and exact solutions*. International Journal of Non-Linear Mechanics 111, p. 28-41 (2019)

[4] I.V. Chernykh, *Modeling of electrical devices in MATLAB, SimPowerSystems and Simulink.* **288** (2007)

[5] O.S. Kozlov, D.E. Kondakov, L.M. Skvortsov, K.A. Timofeev, V.V, Khodakovsky, *The program complex "Modeling in technical devices" (PC "MVTU")*, http://model.exponenta.ru/mvtu/20050615.html

[6] O.S Kozlov, L.M. Skvortsov, *MVTU software package in scientific research and applied developments*. Mathematical Models and Computer Simulations 8(4), p. 358-368 (2016)

[7] G. Mao, L.R. Petzold, *Efficient integration over discontinuities for differential-algebraic*

systems. Computers & Mathematics with Applications, **43**, Issues 1–2, p. 65-79 (January 2002)

[8] L.M. Skvortsov, *The explicit multi-step method for the numerical solution of stiff differential equations* // Journal of Computational Mathematics and Mathematical Physics. T. 47. No. 6. p. 959-967 (2007)

[9] Y.A. Evstifeev, V.B. Manichev. *An effective A-stable method for integrating ordinary differential equations for electronic circuit analysis programs.* Radio electronics. T.29, № 11. p. 31–35 (1986)

[10] D.M. Zhuk, V.B. Manichev, A.O. Ilnitsky, *Methods and algorithms for solving differential-algebraic equations for modeling systems and objects in the time domain.* Information Technologies. №7, part1. p. 16–24 (2010)

[11] L.M. Skvortsov, *Implicit Runge–Kutta Methods with Explicit Internal Stages.* Computational Mathematics and Mathematical Physics 58(3), p. 307-321 (2018)

[12] J.C. Butcher, *Numerical methods for ordinary differential equations.* John Wiley & Sons Ltd, **484** (2008)

[13] D.M. Zhuk, V.B. Manichev, M.K. Sakharov, *SADEL as a library of "ultra-precise" solvers for the PA10 software package (SADEL-PA10).* MES, p. 147-153 (2012)

[14] V.B. Manichev, D.M. Zhuk, F.A. Vityukov, *Method of mathematical testing of the analysis of transient processes in CAD electronic.* MES, p. 166-172 (2014)

[15] V.N. Bakulin, Y.I. Vinogradov, *Analytical and asymptotic solution of boundary value problems in the mechanics of deformed shells under concentrated loading.* Russian Aeronautics 60(1), p. 13-20 (2017)

[16] D. Kahaner, C. Mouler, S. Nash, *Numerical methods and software*, Prentice Hall, **494** (1989)

[17] F. Mazzia, C. Magherini, F. Iavernaro, *Test set for initial value problem solvers.* http://pitagora.dm.uniba.it/~testset

[18] O.S. Kozlov, L.M. Skvortsov, *Test comparison of MATLAB ODE Solvers.* IPU RAS Publishing House, p. 53–60 (2002)

[19] Y.P. Petrov, *Ensuring the reliability and robustness of computer calculations.* BHV-Petersburg, **160** (2008)

[20] E.F. Mitenkova, E.V. Solov`eva, V.B. Manichev, E.O. Fel'dman, *Isotopic Kinetics Problems on a Complete Elemental Basis of the Fission-Product Yield.* Atomic Energy 124(1), p. 69-73 (2018)

IOP Conf. Series: Materials Science and Engineering **630** (2019) 012013 doi:10.1088/1757-899X/630/1/012013

Temperature simulation of system-in-package produced with hybrid chip mounting technology

Petr Boriskov[1], Natalya Ershova[1, a], Vadim Putrolaynen[1], Maxim Savitskii[1,2], Pavel Seredov[1] and Alexei Soloviev[1]

[1]Petrozavodsk State University, 31 Lenina str., Petrozavodsk 185910, Russia
[2]GS Nanotech, 11 Industrialnaya str., Gusev 238050, Russia

[a]ershova@petrsu.ru

Abstract. Finite element computer simulation was used to calculate the temperature distribution of an integrated circuit in the System-in-Package (SiP) produced with hybrid chip 2D mounting technologies such as flip chip and wire bonding. Technical recommendations on the choice of the mold compound are proposed, taking into account the continuous layer of the mold compound separating the active elements and the heat sink radiator.

1. Introduction

Proper thermal conditions during long and continuous operation are important factors affecting the reliability of equipment [1-3]. Heat removal from the most thermally loaded elements of modern semiconductor and hybrid System-in-Package (SiP) is a particularly topical challenge for microelectronics industry. This challenge is affected by many factors, such as reduction of topological size, improvement of the characteristics of dice, method of formation of contacts, placement of elements in the housing, as well as the choice of compound at the stage of SiP molding.

Nowadays SiP is one of the most common housing technology for electronic systems. Developers of SiP pay more attention to the increase of the number of different embedded functions of a device in the most reliable and cheap way than the increase of the number of transistors used. Dice in SiP and their supplemented passive or other necessary components are arranged on the same level (2D integration) or one above the others (3D integration). Thus, full functioning electronic device is formed from integrated modules as a single package.

An additional issue of SiP development is the necessity of combining various mounting technologies. E.g. dice with a large number of pads, such as processors, are mounted using flip chip technology, while others like memory dice are still mounted directly to the substrate using wire bonding. The processor is mounted to the board of such hybrid SiP in an inverted form with the active side down (flip chip). Then the processor die is fixed on the substrate by means of ball contacts. Finally the remaining voids between the board and the die are sealed with underfill. Flash memory dice should be placed on the substrate with the active side up (direct die placement), and their contacts should be attached by wire bonding to the pads of the substrate.

The individual active modules of such a hybrid SiP are heated differently during the operation [3, 4]. A large number of contacts plays a significant role in this difference, because despite the total small area of contacts their thermal conductivity is maximal compared to other materials (silicon and CEM). This issue raises the problem of optimal placement of dice on the board, and also requires selection of a proper mold compound, such as epoxy molding sand [5, 6]. If the developed block does not have high-current circuits, the dielectric strength of the mold compound fades into the background. The most important parameters for the task are a low coefficient of thermal expansion, low shrinkage during polymerization (so as not to damage the tracks when molding the board) and high coefficient of thermal conductivity.

In this work we used the finite element computer simulation of the temperature distribution of the SiP to propose technical recommendations on the choice of the mold compound and the process of its molding, taking into account a system-in-package with a powerful heat-generating element (processor) assembled by means of the hybrid 2D chip packaging technologies, i.e. flip chip and wire bonding.

The remainder of the paper is organized as follows. Description of the problem is given in Section 2. The results of modelling are discussed in Section 3. Section 4 concludes the research.

2. Features of assembly and model scheme of SiP

Figure 1 (a) shows the developed 3D model of the SiP layout, which consists of several heat generation elements, specifically a processor and two memory dice mounted on the PCB. The processor with approx. 3000 ball contacts is mounted using flip chip technology. The memory dice are connected using wire bonding. We use a copper radiator installed on the surface of the mold compound to improve heat exchange with the environment and reduce the temperature of the dice. The surface area of the radiator is 64 cm^2. Our calculation considers the developed SiP as a substrate and 16 conducting layers separated by a dielectric.

Figure 1 shows that the processor protrudes above the memory dies. The processor is the most bulky and heating element, and therefore it has a thinner mold compound layer above than other dice. This imposes certain restrictions, both on the process of pouring the mold compound and on the heat removal from the thermally loaded elements during SiP operation. The fact is that the pouring of the hot compound mixture always starts on one side of the SiP and goes evenly to the opposite side, that's how the entire SiP is molded completely. According to the technological regulations, grinding (thinning) of the mold compound layer after its hardening is not provided, due to the fact that this operation is not workable and may cause the undesirable loss of the overall mechanical strength and sealing of the SiP. Thus, a certain mold compound layer always remains above the processor (Figure 1 (b)). Such layer affects the heat sinking due to its low thermal conductivity (compared to the silicon substrate of the processor and the copper radiator). But thickness of this layer can be varied by the height of the molds for casting mold compound.

We use Cadence Sigrity software package for finite-element method simulation based on the heat equation and the conditions of cross-linking of heat fluxes at the boundaries of regions. Table 1 details parameters of heat generating elements of SiP.

The desired parameters of the problem were the maximal temperatures in the processor and memory dice depending on the thermal conductivity of the mold compound and the thickness of mold compound layer above the processor.

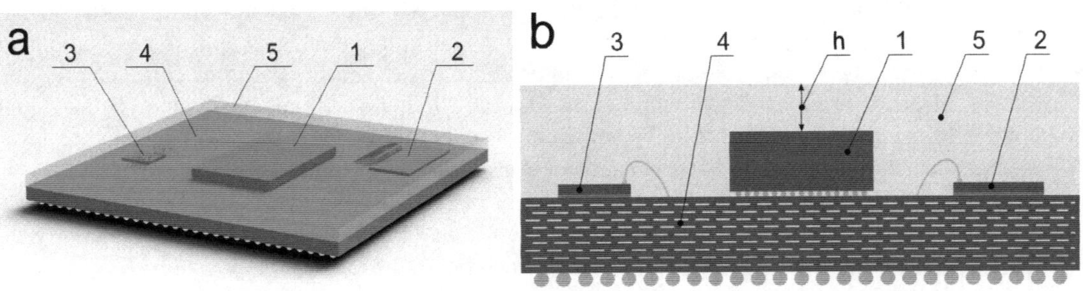

Figure 1. General 3D view (a) and 2D section (b) of 3D SiP layout, 1 – die no. 1 (processor), 2, 3 – dies no. 2 and no. 3 (memory), 4 – multilayer substrate, 5 – mold compound

Table 1. Parameters of heat generating elements of SiP.

Die	Mounting technology	Heat dissipation power, mW	Dimensions of the die, mm
1	flip chip	7000	12.6 x 8 x 0.8
2	wire bonding	82.5	9 x 4.13 x 0.2
3	wire bonding	13.2	2.925x3.025x0.2

3. Results and discussion

All the calculations show that the processor has the highest temperature, because it is the most powerful heat generating element. In our problem we consider the processor with heat dissipation power of 7 W (see Table 1). Processor's temperature exceeds the allowable maximum (Tmax=80 °C) in the absence of the radiator regardless of the layer thickness and thermal conductivity of the mold compound. Therefore presence of the radiator is fundamentally necessary for effective heat removal from the operating SiP. Figure 2 shows an example of the temperature distribution in the SiP, obtained in modelling by Cadence Sigrity.

Figure 2. Temperature distribution in the case of mold compound with thermal conductivity $\lambda \sim 0.5$ W/(m·K) and thickness of mold compound layer h=200 μm.

Further we present the results of temperature simulations at various points of the SiP with a copper radiator over the mold compound layer. During simulations we varied thermal conductivity of the mold compound and thickness of mold compound layer. Figure 3 shows the calculated dependencies of maximal temperatures on the thermal conductivity of the mold compound at layer thicknesses h = 500 and 100 μm.

Figure 3 shows that at the initial stage the processor temperature changes sharply with an increase in the thermal conductivity coefficient of mold compound. With a further increase in λ the change in the temperature becomes flatter. Thus, for the structure under consideration it is possible to choose the value of thermal conductivity, which provides effective reduce of the temperature in the processor. It should be noted that the temperature in the processor depends on the thickness of the mold compound layer that separates the processor from the radiator. The higher the thickness of this layer is, the stronger the dependence of the temperature on compound thermal conductivity is. The minimal temperature in the processor may be achieved in case of the absence of the mold compound layer (h = 0). But this case is not considered because of the technological reasons mentioned above. It is also worth noting that the temperature in the compound surface increases with λ and at certain values exceeds the temperatures of the die no. 1 and the die no. 2. This feature indicates an effective heat sink, i.e. the heat flow is directed towards the radiator and not at dice in the SiP.

Figure 4 (a) shows graphs of the temperature change rate in the processor depending on the compound thermal conductivity and the thickness of the mold compound layer above the processor. Decrease in the temperature change rate drops sharply starting with $\lambda \sim 0.4$ W/(m·K).

On the one hand, Figure 4 (a) does not show a significant change in the temperature of the processor for different thicknesses of the mold compound layer. However, closer examination of Figure 4 (b) shows that maximal processor temperature decreases significantly at thicknesses less than 400 μm. Thus, there is an optimal range of mold compound parameters (thermal conductivity coefficient and thickness of mold compound layer above the processor). These parameters determine the required temperature regime of dice in the SiP.

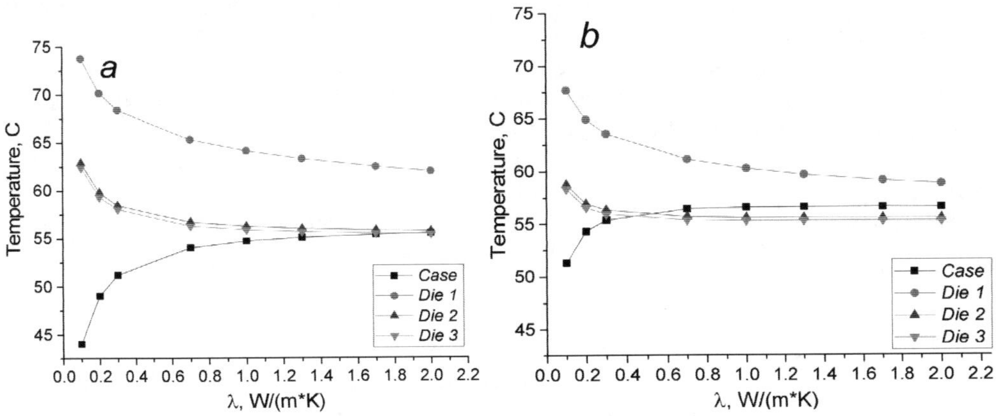

Figure 3. Dependency of maximal temperatures in the surface of mold compound (T_{Case}), the processor (T_{Die1}) and the memory dice (T_{Die2} and T_{Die3}) on compound thermal conductivity at thicknesses of mold compound layer h=500 μm (a) and h=100 μm (b).

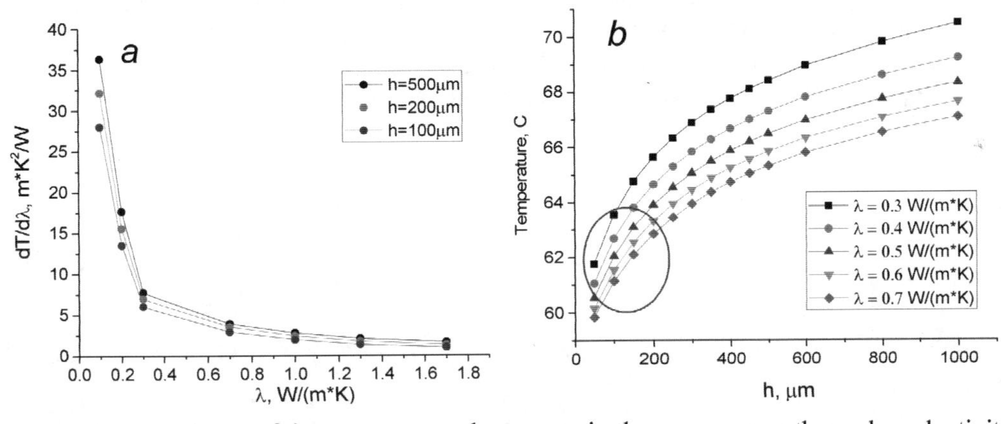

Figure 4. Dependency of the temperature change rate in the processor on thermal conductivity (dT/dλ) (a), and dependency of the temperature in the processor (T_{Die1}) on the thickness of mold compound layer at various values of compound thermal conductivity (b)

The choice of mold compound is also determined by the technology of filling and its price. The excess of the mold compound layer of more than 500 μm thickness over the processor is impractical because of additional consumption of mold compound. This means increase in the cost per SiP with excessive strength of the protective layer. A compound with higher thermal conductivity costs more expensive, since its synthesis requires additives from heat-conducting materials [5, 6]. In addition, mold compound with higher thermal conductivity dissipates heat more efficiently when pouring over the SiP (as a hot mixture), i.e. it hardens faster and that can also lead to certain problems in this process.

In our opinion, the best option for selecting a mold compound is the one, in which maximal processor temperature does not exceed 65 °C (for using SiP in rooms without ventilations). In this case the mold compound may have a thermal conductivity in the range of 0.4÷0.7 W/(m·K), e.g. a common mold compound with λ ~ 0.6 W/(m·K). It should be poured over the processor with a relatively thin layer of approx. 100÷200 μm. In the range of these parameters (Figure 4 (b), highlighted by a circle) the temperature change due to thermal conductivity (dT/dλ) drops sharply as shown in Figure 4 (a), that gives additional advantages in the variation of the mold compound parameters during its selection.

4 Conclusion

In this paper we took into account position of dies and their mounting method (flip chip, wire bonding) in order to perform temperature simulation of the SiP and then we provide recommendations on the choice of mold compound.

In conclusion, we note that the presented results may be useful for the technical regulation of the process of pouring the mold compound not only for this SiP, but also for other microcircuits created using 2D packaging technology System-in-Package.

Acknowledgment

This work was supported by the Ministry of Science and Higher Education of the Russian Federation (project no. 05.577.21.0293, unique identifier RFMEFI57718X0293).

References

[1] B. S. Lall, B. M. Guenin, R. J. Molnar, Proceedings of 1995 IEEE/CPMT 11th Semiconductor Thermal Measurement and Management Symposium, 72–79 (1995).

[2] B. A. Zahn, *IEEE Transactions on Components and Packaging Technologies*, **23** (1), 33–39 (2000)

[3] I. Yunhyeok [et al.], Twentieth Annual IEEE Semiconductor Thermal Measurement and Management Symposium, 5 (2004)

[4] C.-K. Liu [et al.], 3rd International Microsystems, Packaging, Assembly & Circuits Technology Conference, 379–382 (2008)

[5] S.Komori, Y. Sakamoto, Materials for Advanced Packaging, 339–363 (2009)

[6] A. Mavinkurve, L. Goumans, J. Martens, European Microelectronics Packaging Conference (EMPC), 7 (2013)

A Miniature Study of Directional Couplers, Each with a Different Central Frequency

Pham Huu Tan[1], Nguyen Khac Khiem[1], Luu Quang Hung[1,2] and Nikolay S. Knyazev[2]

1 Vietnam Maritime University, Hai Phong, Vietnam
2 Ural Federal University, Yekaterinburg, Russian

phamhuutan@vimaru.edu.vn, nkk@vimaru.edu.vn, hung.luuquang@vimaru.edu.vn, nknyazev@yandex.ru

Abstract. A study to determine how the operating frequency of the device during miniaturization depends on its efficiency and the safety of the original characteristics. It was found that the miniaturization efficiency decreases with increasing frequency. Thus, the design of compact couplers for Central frequencies from 1000 to 3000 MHz with a step of 500 MHz was studied. In this case, the dimensions in relation to the standard design were reduced by 78.8, 62.5, 46.7, 39 and 30%, respectively.

1. Introduction

The operating frequency of directional couplers affects their overall dimensions, the lower it is the larger the area occupied by the couplers. Therefore, at low frequencies it is necessary to apply miniaturization of all structural elements to obtain a device with acceptable dimensions. Miniaturization is to replace the standard segment with compact structure. In this case, these structures must have similar characteristics for the safety of the characteristics of the branch. It is also necessary to take into account how the Central frequency affects the effectiveness of this approach to reduce the geometric dimensions of the branch. In order to evaluate the existing experience of researchers in miniaturization, some articles were considered. In [1] proposed to reduce the size using quasi-concentrated elements, in [2] equivalent transmission lines, in [3] U-shaped capacitances, in [4] periodic capacitive loads, asymmetric T-shaped structures in [5], low-pass filters in [6,7,8], slow-down systems in [9,10], artificial transmission lines in [11-13], fractal structures in [14,15], in [16] high-resistance elements, in [17] loaded loops, in [18-21] interdigital capacitors. In this paper, we will investigate how the efficiency of the coupler at different Central frequencies changes with the same technique.

2. Design

Initially, the procedure was carried out to design standard designs of couplers at Central frequencies from 1000 to 3000 MHz in increments of 500 MHz. For all structures the same microwave material was used, with a permittivity of 4.4, a thickness of 1 mm and losses in the dielectric of 0.02. Figure 1 shows the topology of the coupler operating at a Central frequency of 1 GHz, and its frequency characteristics are shown in figure 2. The area of such a device is 2060 mm^2. The operating frequency band measured at the level of junction 20 dB is 110 MHz.

IOP Conf. Series: Materials Science and Engineering **630** (2019) 012014 doi:10.1088/1757-899X/630/1/012014

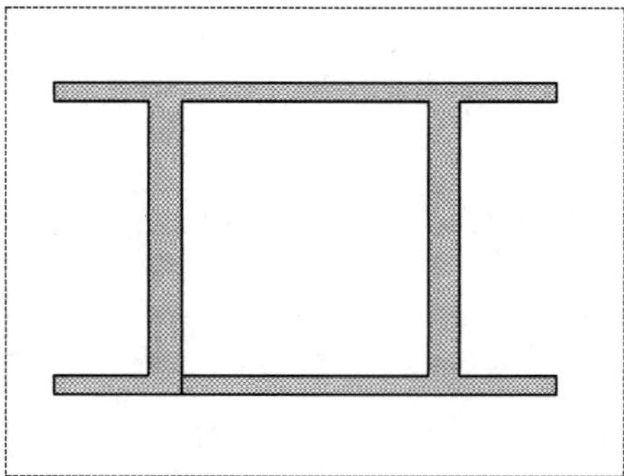

Figure 1. Topology directional coupler

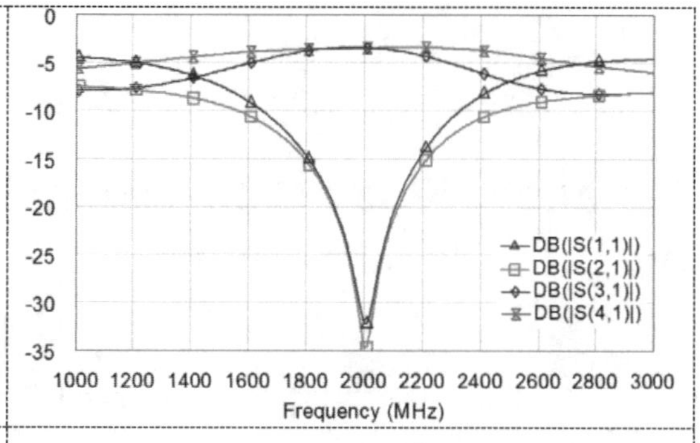

Figure 2. S-parameters from frequency

Standard designs were synthesized for all Central frequencies of 1000,1500,2000,2500 and 3000 MHz. After that, for all quarter-wave segments, the topologies of compact structures were calculated, which will be installed in place of standard segments. An example of such a structure is shown in figure 3. The frequency characteristics of such a structure are also presented in order to analyze its similarity with the characteristics of the quarter-wave segment (Fig. 4, 5).

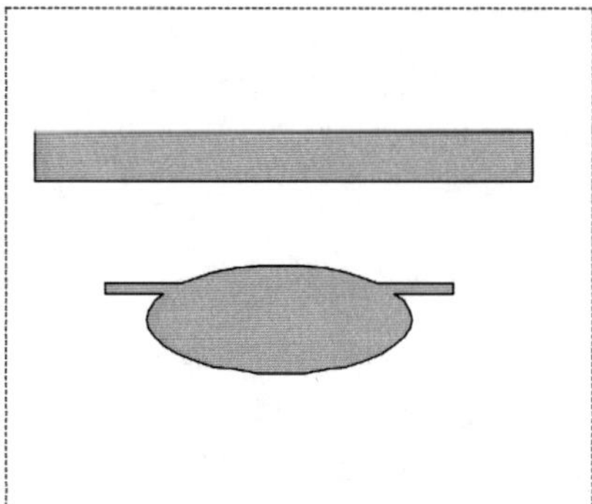

Figure 3. Topology of a quarter-wave segment and a compact structure with an impedance of 50 Ohm

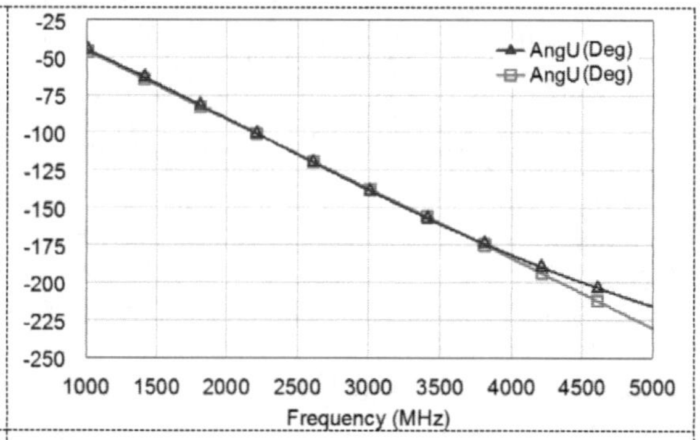

Figure 4. Phase on frequenc

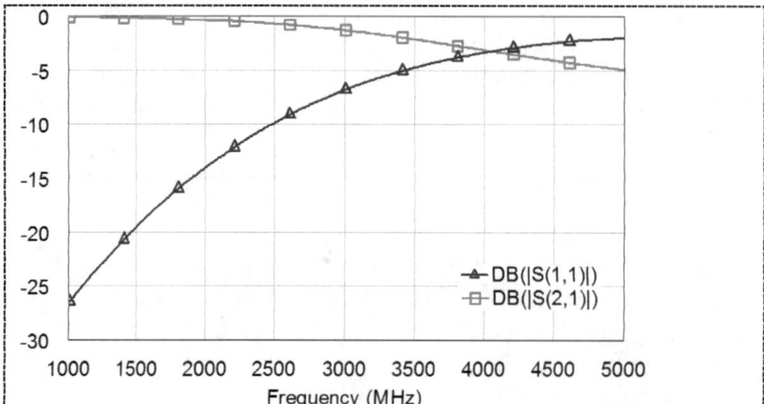

Figure 5. Frequency characteristics of a compact structure

It can be seen that in the required frequency range, for the coupler at the Central frequency of 1000 MHz, the structure coincides with the characteristics of the segment. After replacing all segments with compact structures, we obtained a compact design of the coupler at a Central frequency of 1000 MHz. The area of such a device is 435 mm^2, and the operating frequency band is equal to the value of 61 MHz. The topology is shown in figure 6 and the characteristics in figure 7.

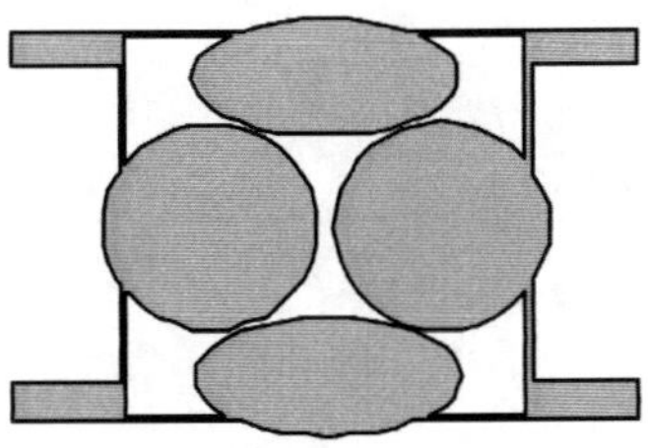

Figure 6. Compact coupler topology

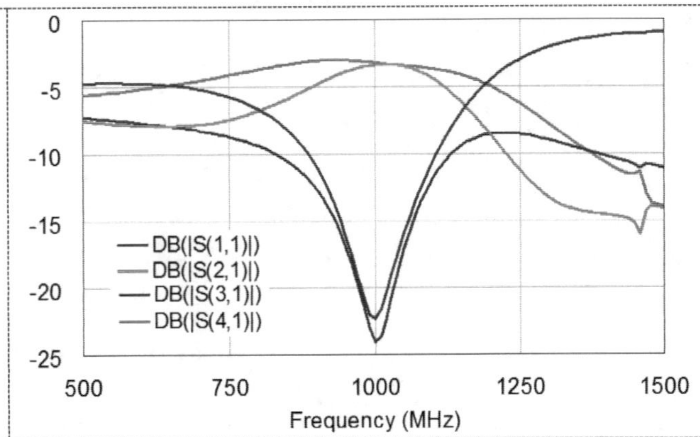

Figure 7. S-parameters from frequency

Compact coupler for a center frequency of 1500 MHz. The area of such a device is 376 mm2, and the operating frequency band is equal to the value of 121 MHz. The topology of the device is shown in figure 8, and its frequency characteristics are illustrated in figure 9.

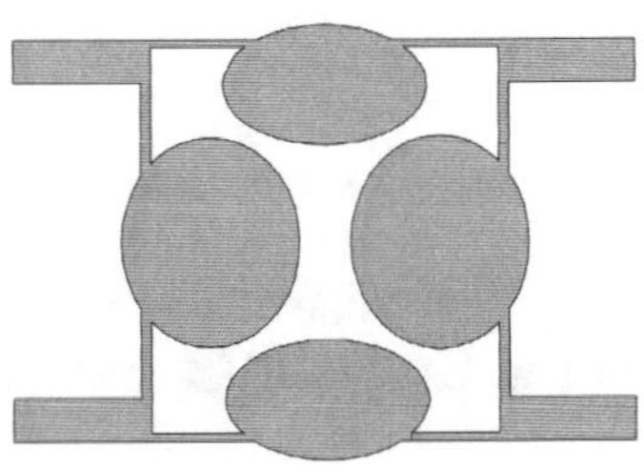

Figure 8. Compact coupler topology

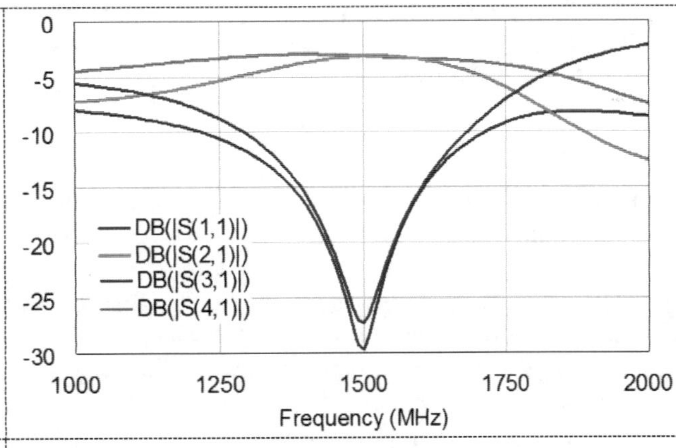

Figure 9. S-parameters from frequency

Compact coupler for a center frequency of 2000 MHz. The area of such a device is 332.9 mm2, and the operating frequency band is equal to the value of 160 MHz. The topology of the device is shown in figure 10, and its frequency characteristics are illustrated in figure 11.

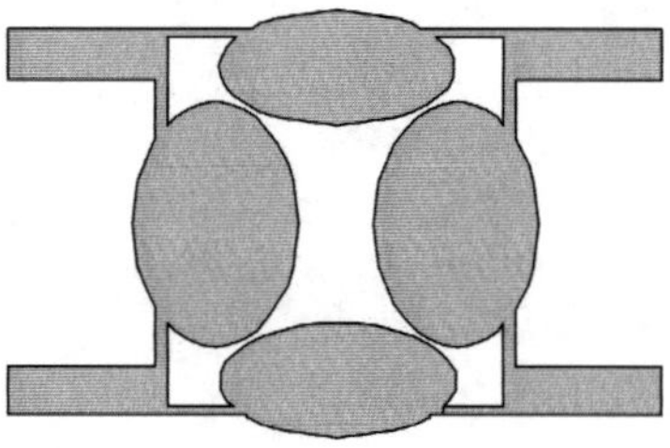

Figure 10. Compact coupler topology

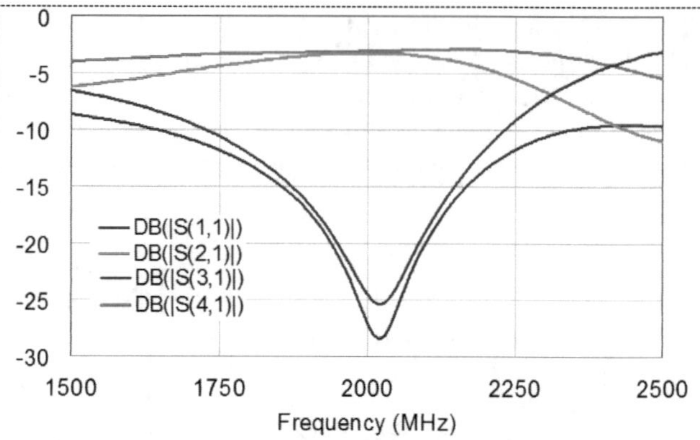

Figure 11. S-parameters from frequency

Compact coupler for a center frequency of 2500 MHz. The area of such a device is 265.3 mm2, and the operating frequency band is equal to the value of 195 MHz. The topology of the device is shown in figure 12, and its frequency characteristics are illustrated in figure 13.

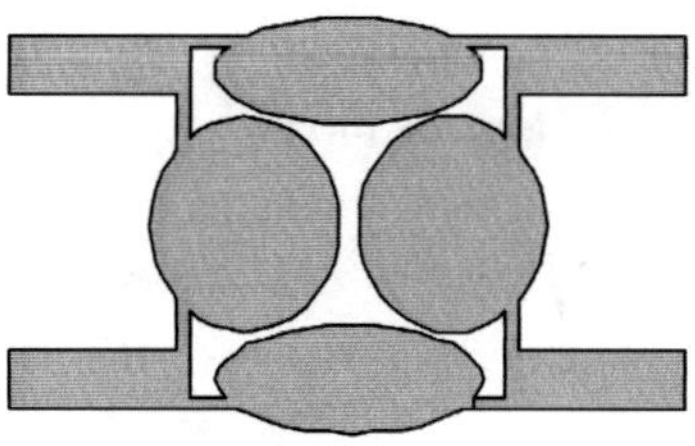

Figure 12. Compact coupler topology

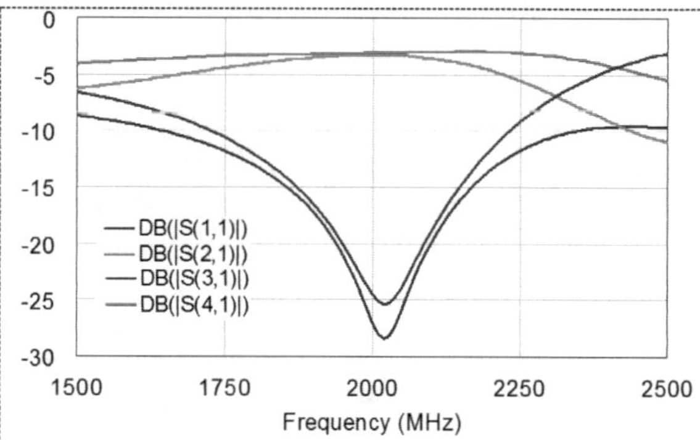

Figure 13. S-parameters from frequency

Compact coupler for a center frequency of 3000 MHz. The area of such a device is 229.46 mm2, and the operating frequency band is equal to the value of 234 MHz. The topology of the device is shown in figure 14, and its frequency characteristics are illustrated in figure 15. For comparison, the results were listed in table 1.

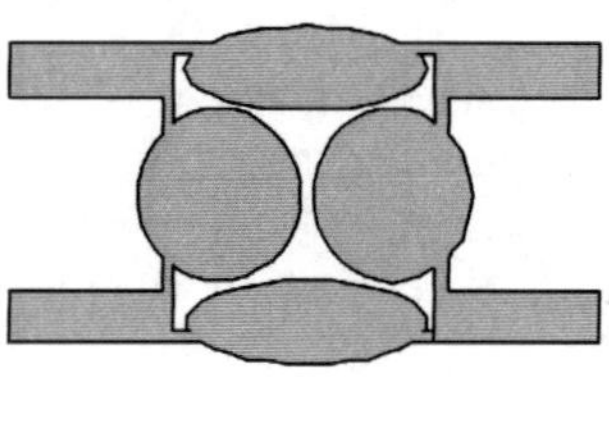

Figure 14. Compact coupler topology

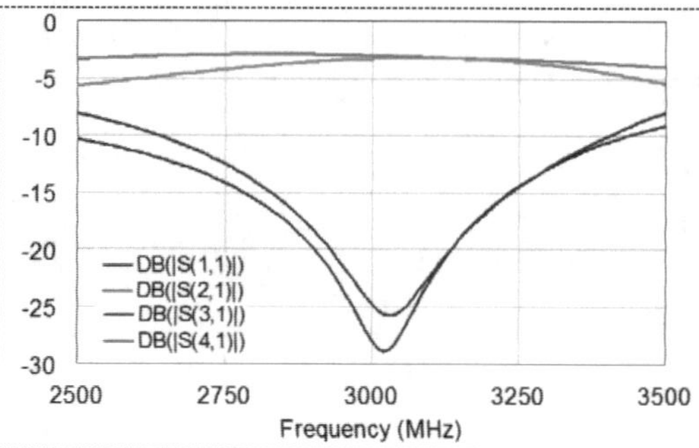

Figure 15. S-parameters from frequency

Table 1. Comparison of coupler

Frequency	Design	Area, mm2	Bandwidth, MHz	Reduce area, %
1000 MHz	Standard	2060	110	78.8%
	Compact	435	61	
1500 MHz	Standard	1005	168	62.5%
	Compact	376	121	
2000 MHz	Standard	625	224	46.7%
	Compact	332.9	160	
2500 MHz	Standard	435	273	39%
	Compact	265.3	195	
3000 MHz	Standard	324	321	30%

3. Conclusion

As a result of the study of how the Central frequency of the coupler affects the efficiency of the miniaturization method, it was found that with an increase in the Central frequency, the capabilities of the proposed miniaturization method decrease, since the area of the device at the Central frequency of 1000 MHz was reduced by 78.8%, and at a frequency of 3000 MHz by only 30%.

4. References

[1] S.-S. Liao and J.-T. Peng, "Compact planar microstrip branch-line couplers using the quasi-lumped elements approach with nonsymmetrical and symmetrical T-shaped structure," *IEEE Trans. Microw. TheoryTech., vol. 54, pp. 3508-3514, Sep. 2006.*

[2] Letavin Dènis, "Miniature microstrip branch line coupler with folded artificial transmission lines," *AEU - International Journal of Electronics and Communications, vol. 99, pp. 8-13, Feb. 2019. DOI: 10.1016/j.aeue.2018.11.016.*

[3] Letavin Denis A., "Compact branch-line coupler based on an inductor and a u-shaped capacitor," *19th International Conference of Young Specialists on Micro/Nanotechnologies and Electron Devices, EDM 2018, pp. 195-198, July 2018. DOI: 10.1109/EDM.2018.8435096.*

[4] K. W. Eccleston, S. H. M. Ong, "Compact planar microstripline branch-line and rat-race coupler," *IEEE Trans. Microw. Theory Tech., vol. 51, pp. 2119-2125, Oct. 2003.*

[5] S.-S. Liao, P.-T.Sun, N.-C.Chin and J.-T.Peng, "A novel compact-size branch-line coupler," *IEEE Microw.Wireless Compon.Lett. vol. 15, pp. 588-590, Sep. 2005.*

[6] Letavin Denis A., "Compact crossover based on low-pass filters," *19th International Conference of Young Specialists on Micro/Nanotechnologies and Electron Devices, EDM 2018, pp. 192-194, July 2018. DOI: 10.1109/EDM.2018.8434978.*

[7] Letavin Denis A., "Compact microstrip three-loop coupler," *2017 International Applied Computational Electromagnetics Society Symposium - Italy, ACES 2017, May. 2017, DOI: 10.23919/ROPACES.2017.7916036.*

[8] Letavin Denis A., "Miniature branch-line coupler structure analysis," *18th International Conference of Young Specialists on Micro/Nanotechnologies and Electron Devices, EDM 2017, July. 2017, pp. 99-101. DOI: 10.1109/EDM.2017.7981717.*

[9] J. Wang, B. Z. Wang, Y.-X.Guo, L.-C.Ong, and S. Xiao "A compact slow-wave microstrip branch-line coupler with high performance," *IEEE Microw.Wireless Compon.Lett. vol. 17, pp. 501-503, Jul. 2007.*

[10] Wei-Shin Chang and Chi-Yang Chang, "A High Slow-Wave Factor Microstrip Structure With Simple Design Formulas and Its Application to Microewave Circuit Design," *IEEE Trans. Microw. Theory Tech., vol. 60, no. 11, pp. 3376-3383, Nov. 2012.*

[11] Letavin Denis A., "Compact dual-frequency microstrip branch-line coupler using artificial transmission lines," *19th International Conference of Young Specialists on*

Micro/Nanotechnologies and Electron Devices, EDM 2018, pp. 185-188, July 2018. DOI: 10.1109/EDM.2018.8434988.

[12] Letavin Denis A., "Two Methods for Miniaturization of Stub Quadrature Couplers," *Journal of Communications Technology and Electronics, vol. 63, Issue 8, Aug. 2018, pp. 933-935. DOI: 10.1134/S1064226918080107*

[13] Letavin Denis A., "Five-Stub Directional Device with Substantially Reduced Overall Dimensions," *2018 IEEE Radio and Antenna Days of the Indian Ocean, RADIO 2018. DOI: 10.23919/RADIO.2018.8572386*

[14] H.Ghali, T. A. Moselhy, "Miniaturized fractal rat-race, branch-line, and coupler-line hybrids," *IEEE Trans. Microw. Theory Tech., vol. 52, pp. 2513-2520, Nov. 2004.*

[15] J. Zhu, Y. Zhou, J. Liu, "Miniaturization Of Broadband 3-Db Branch-Line Coupler," *Progress In Electromagnetics Research Letters, vol. 24, pp. 169-176, 2011.*

[16] C. W. Tang, M.-G.Chen, "Synthesizing microstrip branch-line coupler with predetermined compact size and bandwidth," *IEEE Trans. Microw. Theory Tech., vol. 55, pp. 1926-1934, Sep. 2007.*

[17] Kimberley W. Eccleston, and Sebastian H.M. Ong, "Compact Planar Microstripline Branch-Line and Rat-Race Couplers," *IEEE Trans. Microw. Theory Tech., vol. 51, no. 10, pp. 2119-2125, Oct. 2003.*

[18] Kai-Yu Tsai, Hao-Shun Yang, Jau-Horng Chen, and Yi-Jan Emery Chen, "A miniaturized 2 dB Branch-Line Hybrid Coupler With Harmonics Suppression," *IEEE Microw. Wireless Compon. Lett., vol. 21, no. 10, pp. 537-539, Oct. 2011.*

[19] Letavin Denis A., "Compact directional coupler with five stubs," *2018 IEEE Radio and Antenna Days of the Indian Ocean, RADIO 2018. DOI: 10.1088/1757-899X/524/1/012007*

[20] Ashmi Chakraborty Das, Lakhindar Murmu, Santanu Dwari "A Compact Branch-Line Coupler using Folded Microstrip Lines," *2013 International Conference on Microwave and Photonics (ICMAP).*

[21] Kae-Oh Sun, Sung-Jin Ho, Chih-Chuan Yen, and Daniel van der Weide, "A Compact Branch-Line Coupler Using Discontinuous Microstrip Lines," *IEEE MICROWAVE AND WIRELESS COMPONENTS LETTERS, VOL. 15, NO. 8, AUGUST 2005*

Miniature directional coupler with different dielectric parameters

Luu Quang Hung[1,2], Pham Huu Tan[1] and Nikolay S. Knyazev[2]

1 Vietnam Maritime University, Hai Phong, Vietnam
2 Ural Federal University, Yekaterinburg, Russian

hung.luuquang@vimaru.edu.vn, phamhuutan@vimaru.edu.vn, nknyazev@yandex.ru

Abstract. A study aimed at identifying the effect of substrate parameters such as dielectric permittivity and thickness on the characteristics of directional couplers, whose dimensions are reduced by means of compact structures. It was found that with the increase in the thickness of the substrate in a compact coupler increases the frequency band and the degree of miniaturization, and with increasing dielectric constant Vice versa.

1. Introduction

In the development of directional couplers it is important to consider what is the impact of a particular parameter of the microwave substrate on its characteristics. This is especially important in the design of compact couplers, since in this case the substrate material can affect the degree of miniaturization, frequency band, and loss. Various compact structure applied with the purpose to reduce the dimensions of the microstrip taps with a minimum deterioration in its characteristics. Compact structures consist of a set of inductances and capacitances. The literature has already described many works concerning the miniaturization of taps, consider only some of them, in [1] proposed to reduce the size using quasi-concentrated elements, in [2] equivalent transmission lines, in [3] U-shaped capacitances, in [4] periodic capacitive loads, asymmetric T-shaped structures in [5], low-pass filters in [6,7,8], slow-down systems in [9,10], artificial transmission lines in [11-13], fractal structures in [14,15], in [16] high-resistance elements, in [17] loaded loops, in [18-21] interdigital capacitors. These works and the structures described in them are made on a variety of dielectrics. Therefore, in this paper we will investigate the influence of parameters such as the dielectric constant and the thickness of the microwave dielectric.

2. Design

The study was carried out on a standard model of a coupler with two loops. Its topology operating at a frequency of 2 GHz and calculated in a specialized program is shown in Fig.1. The frequency characteristics of the coupler modeled on a substrate with a dielectric constant of 4.4 and a thickness of 1 mm are shown in Fig.2.3.

IOP Conf. Series: Materials Science and Engineering **630** (2019) 012015 doi:10.1088/1757-899X/630/1/012015

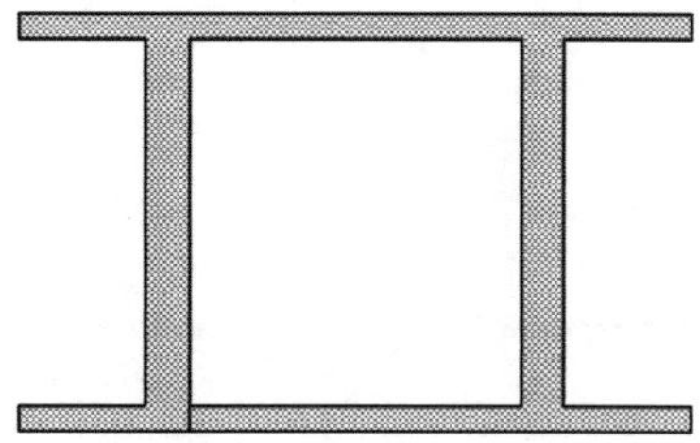

Figure 1. Topology directional coupler

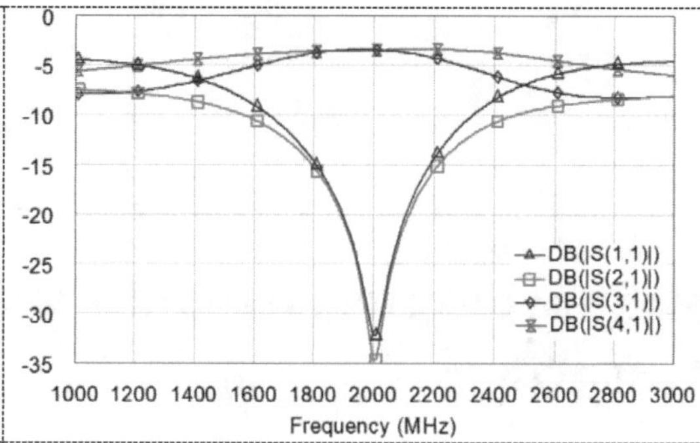

Figure 2. S-parameters from frequency

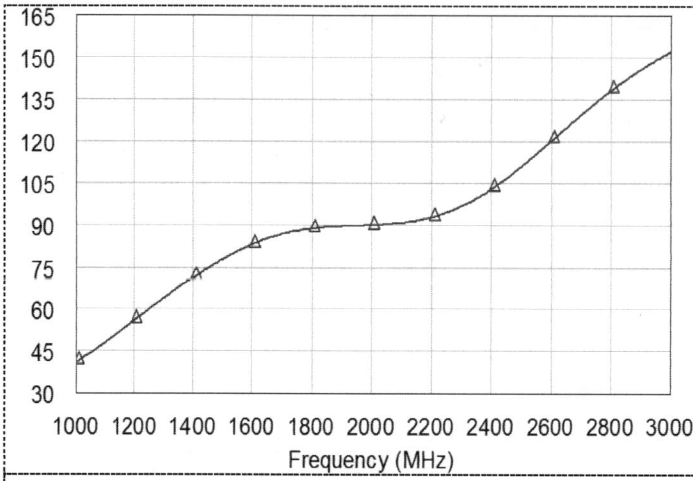

Figure 3. Phase on frequency

The same topology and characteristics were obtained for the dielectric constant of 2.5, 3.4, 4.4 and 5.5 for the thickness of 1 mm. And then with dielectric constant 4.4 with thickness of the substrate 1, 1.5 and 2 mm. This was done to obtain data for later comparison with compact taps. For Fig.4 shows the compact structure and a quarter wavelength segment, and figures 5, 6 show their characteristics.

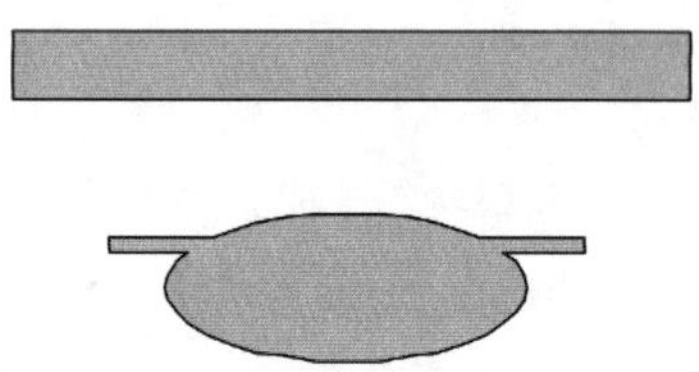

Figure 4. Topology of a quarter-wave segment and a compact structure with an impedance of 50 Ohm

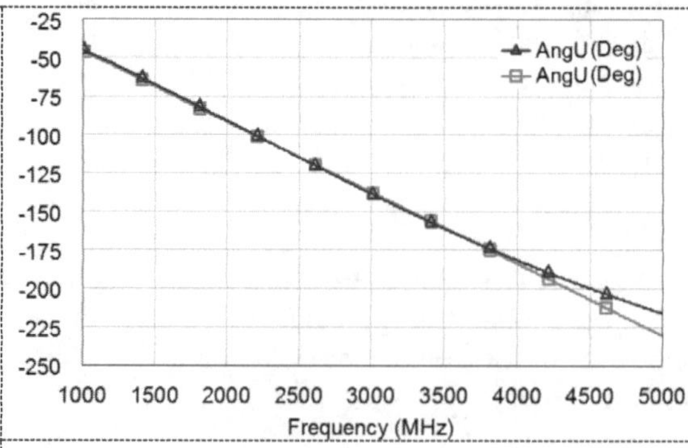

Figure 5. Phase on frequenc

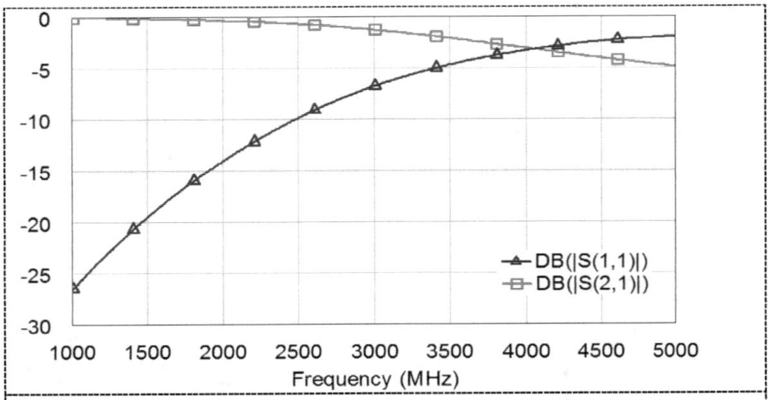

Figure 6. Frequency characteristics of a compact structure

The compact coupler on the substrate e=5.5 and h=1 mm is shown in figure 7, and its frequency characteristics are illustrated in Fig.8.

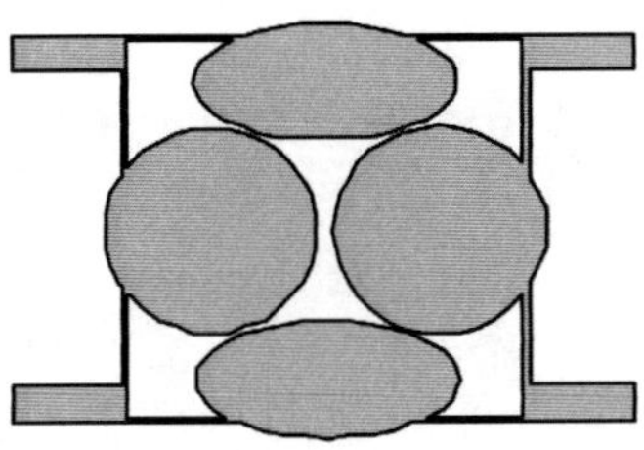

Figure 7. Compact coupler topology

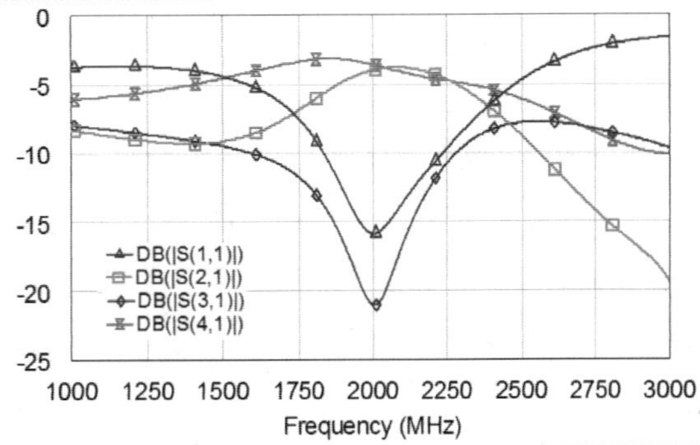

Figure 8. S-parameters from frequency

The compact coupler on the substrate e=4.4 and h=1 mm is shown in figure 9, and its frequency characteristics are illustrated in Fig.10.

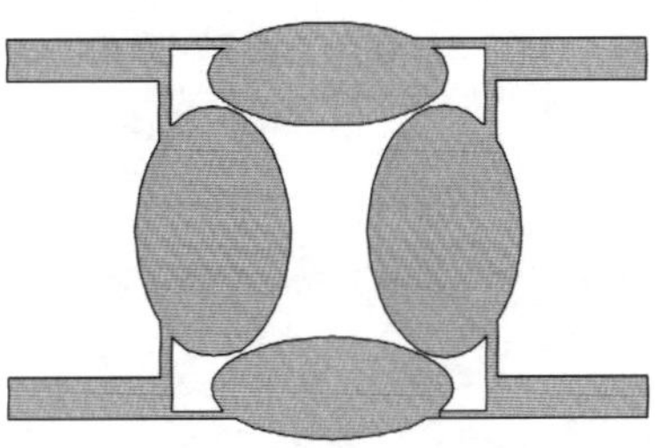

Figure 9. Compact coupler topology

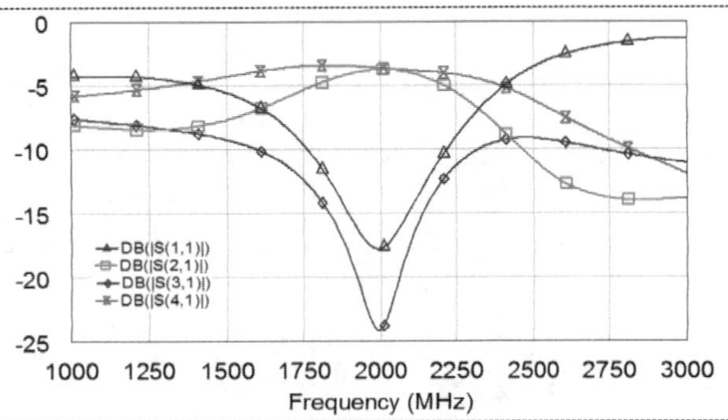

Figure 10. S-parameters from frequency

The compact coupler on the substrate e=3.4 and h=1 mm is shown in figure 11, and its frequency characteristics are illustrated in Fig.12.

IOP Conf. Series: Materials Science and Engineering **630** (2019) 012015 doi:10.1088/1757-899X/630/1/012015

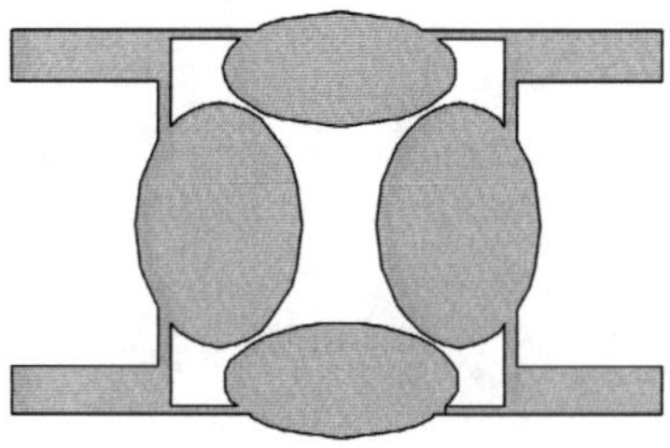

Figure 11. Compact coupler topology

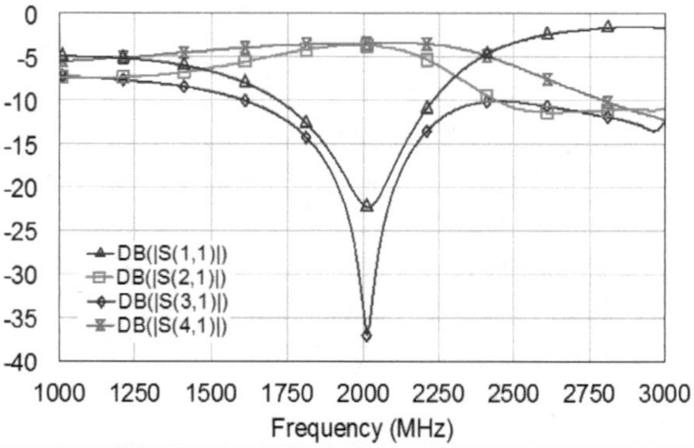

Figure 12. S-parameters from frequency

The compact coupler on the substrate e=2.5 and h=1 mm is shown in figure 13, and its frequency characteristics are illustrated in Fig.14.

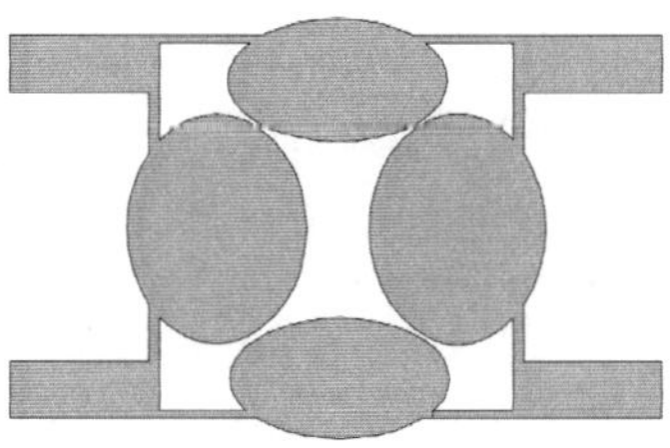

Figure 13. Compact coupler topology

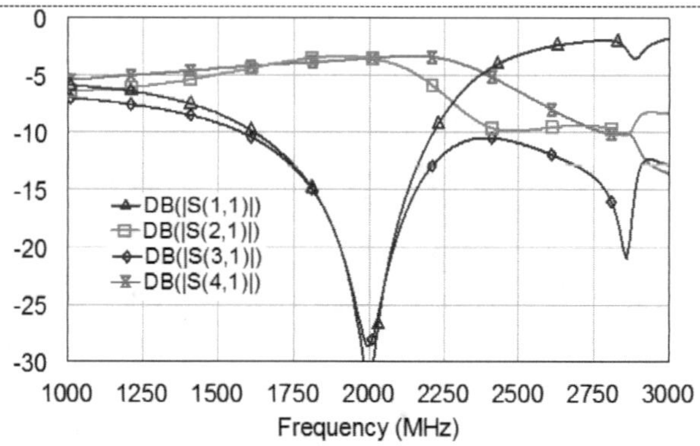

Figure 14. S-parameters from frequency

The compact coupler on the substrate e=4.4 and h=1.5 mm is shown in figure 15, and its frequency characteristics are illustrated in Fig.16.

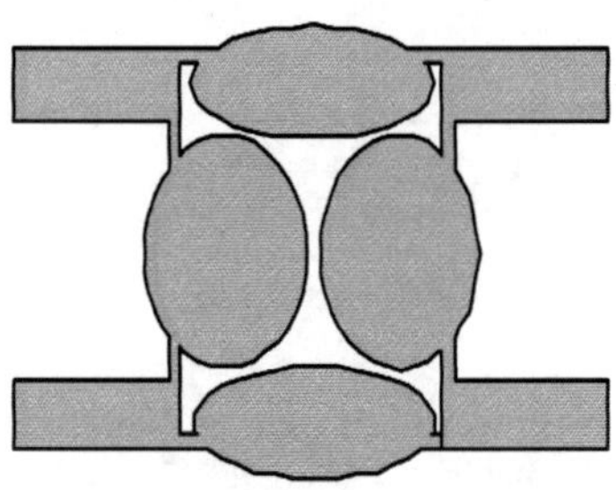

Figure 15. Compact coupler topology

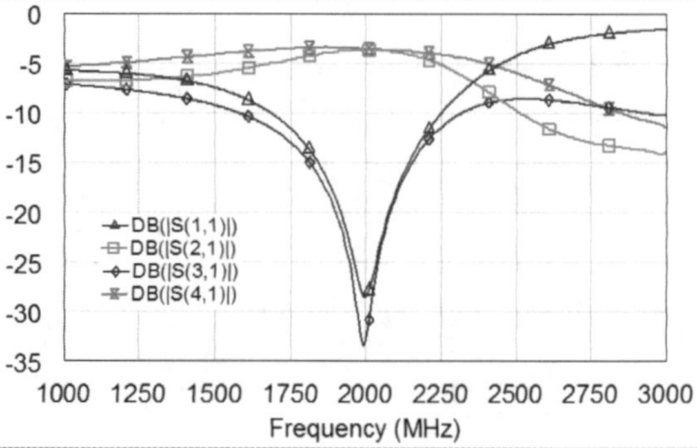

Figure 16. S-parameters from frequency

The compact coupler on the substrate e=4.4 and h=2 mm is shown in figure 17, and its frequency characteristics are illustrated in Fig.18. All the results were listed in table 1.

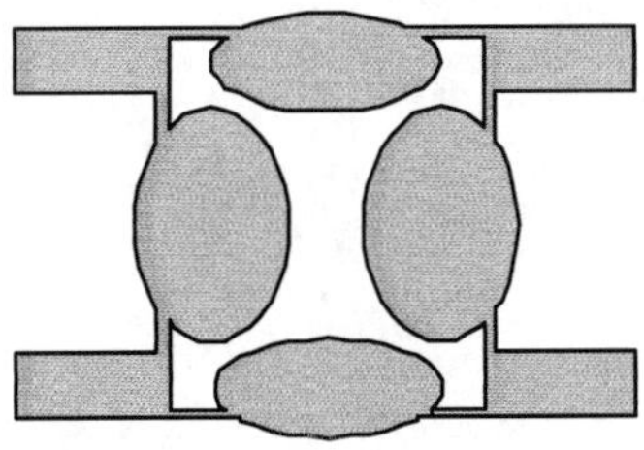

Figure 17. Compact coupler topology

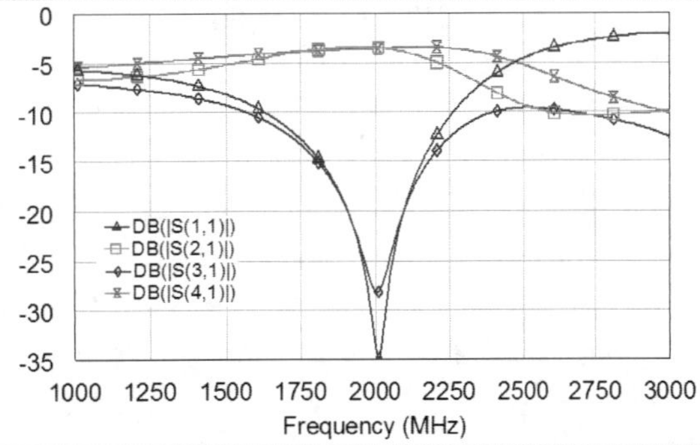

Figure 18. S-parameters from frequency

Table 1. Comparison of coupler

Substrate	Design	Area, mm2	Bandwidth, MHz	Reduce area, % / Reduce bandwidth %
E =2.5	Standard	1037	226	58%
H=1mm	Compact	435	176	22%
E =3.4	Standard	786	225	52.2%
H=1mm	Compact	376	186	17.3%
E =4.4	Standard	628	225	47%
H=1mm	Compact	332.9	140	37%
E =5.5	Standard	512	224	48%
H=1mm	Compact	265.3	80	64%
E =4.4	Standard	753	220	49.5%
H=1.5mm	Compact	380.65	177	19.5%
E =4.4	Standard	875	215	56.5%
H=2mm	Compact	381.5	194	9.7%

3. Conclusion

As a result of the study, it was found that the degree of miniaturization is affected by the dielectric constant and thickness. So the thicker the substrate, the higher the percentage of miniaturization of the design area, as well as in a wider frequency band, the operating frequency band remains at the level of 20 dB decoupling. With increasing dielectric constant miniaturization efficiency decreases as the area decreases, and to preserve the characteristics of the device.

References

[1] S.-S. Liao and J.-T. Peng, "Compact planar microstrip branch-line couplers using the quasi-lumped elements approach with nonsymmetrical and symmetrical T-shaped structure," *IEEE Trans. Microw. TheoryTech., vol. 54, pp. 3508-3514, Sep. 2006.*

[2] Letavin Denis, "Miniature microstrip branch line coupler with folded artificial transmission lines," *AEU - International Journal of Electronics and Communications, vol. 99, pp. 8-13, Feb. 2019. DOI: 10.1016/j.aeue.2018.11.016.*

[3] Letavin Denis A., "Compact branch-line coupler based on an inductor and a u-shaped capacitor," *19th International Conference of Young Specialists on Micro/Nanotechnologies and Electron Devices, EDM 2018, pp. 195-198, July 2018. DOI: 10.1109/EDM.2018.8435096.*

[4] K. W. Eccleston, S. H. M. Ong, "Compact planar microstripline branch-line and rat-race coupler," *IEEE Trans. Microw. Theory Tech., vol. 51, pp. 2119-2125, Oct. 2003.*

[5] S.-S. Liao, P.-T.Sun, N.-C.Chin and J.-T.Peng, "A novel compact-size branch-line coupler," *IEEE Microw.Wireless Compon.Lett. vol. 15, pp. 588-590, Sep. 2005.*

[6] Letavin Denis A., "Compact crossover based on low-pass filters," *19th International Conference of Young Specialists on Micro/Nanotechnologies and Electron Devices, EDM 2018, pp. 192-194, July 2018. DOI: 10.1109/EDM.2018.8434978.*

[7] Letavin Denis A., "Compact microstrip three-loop coupler," *2017 International Applied Computational Electromagnetics Society Symposium - Italy, ACES 2017, May. 2017, DOI: 10.23919/ROPACES.2017.7916036.*

[8] Letavin Denis A., "Miniature branch-line coupler structure analysis," *18th International Conference of Young Specialists on Micro/Nanotechnologies and Electron Devices, EDM 2017, July. 2017, pp. 99-101. DOI: 10.1109/EDM.2017.7981717.*

[9] J. Wang, B. Z. Wang, Y.-X.Guo, L.-C.Ong, and S. Xiao "A compact slow-wave microstrip branch-line coupler with high performance," *IEEE Microw.Wireless Compon.Lett. vol. 17, pp. 501-503, Jul. 2007.*

[10] Wei-Shin Chang and Chi-Yang Chang, "A High Slow-Wave Factor Microstrip Structure With Simple Design Formulas and Its Application to Microewave Circuit Design," *IEEE Trans. Microw. Theory Tech., vol. 60, no. 11, pp. 3376-3383, Nov. 2012.*

[11] Letavin Denis A., "Compact dual-frequency microstrip branch-line coupler using artificial transmission lines," *19th International Conference of Young Specialists on Micro/Nanotechnologies and Electron Devices, EDM 2018, pp. 185-188, July 2018. DOI: 10.1109/EDM.2018.8434988.*

[12] Letavin Denis A., "Two Methods for Miniaturization of Stub Quadrature Couplers," *Journal of Communications Technology and Electronics, vol. 63, Issue 8, Aug. 2018, pp. 933-935. DOI: 10.1134/S1064226918080107*

[13] Letavin Denis A., "Five-Stub Directional Device with Substantially Reduced Overall Dimensions," *2018 IEEE Radio and Antenna Days of the Indian Ocean, RADIO 2018. DOI: 10.23919/RADIO.2018.8572386*

[14] H.Ghali, T. A. Moselhy, "Miniaturized fractal rat-race, branch-line, and coupler-line hybrids," *IEEE Trans. Microw. Theory Tech., vol. 52, pp. 2513-2520, Nov. 2004.*

[15] J. Zhu, Y. Zhou, J. Liu, "Miniaturization Of Broadband 3-Db Branch-Line Coupler," *Progress In Electromagnetics Research Letters, vol. 24, pp. 169-176, 2011.*

[16] C. W. Tang, M.-G.Chen, "Synthesizing microstrip branch-line coupler with predetermined compact size and bandwidth," *IEEE Trans. Microw. Theory Tech., vol. 55, pp. 1926-1934, Sep. 2007.*

[17] Kimberley W. Eccleston, and Sebastian H.M. Ong, "Compact Planar Microstripline Branch-Line and Rat-Race Couplers," *IEEE Trans. Microw. Theory Tech., vol. 51, no. 10, pp. 2119-2125, Oct. 2003.*

[18] Kai-Yu Tsai, Hao-Shun Yang, Jau-Horng Chen, and Yi-Jan Emery Chen, "A miniaturized 2 dB Branch-Line Hybrid Coupler With Harmonics Suppression," *IEEE Microw. Wireless Compon. Lett., vol. 21, no. 10, pp. 537-539, Oct. 2011.*

[19] Ashmi Chakraborty Das, Lakhindar Murmu, Santanu Dwari "A Compact Branch-Line Coupler using Folded Microstrip Lines," *2013 International Conference on Microwave and Photonics (ICMAP).*

[19] Letavin Denis A., "Compact directional coupler with five stubs," *2018 IEEE Radio and Antenna Days of the Indian Ocean, RADIO 2018. DOI: 10.1088/1757-899X/524/1/012007*

[20] Kae-Oh Sun, Sung-Jin Ho, Chih-Chuan Yen, and Daniel van der Weide, "A Compact Branch-Line Coupler Using Discontinuous Microstrip Lines," *IEEE MICROWAVE AND WIRELESS COMPONENTS LETTERS, VOL. 15, NO. 8, AUGUST 2005*

Miniature Directional Coupler with Two Operating Frequency Bands

Nguyen Khac Khiem[1], Luu Quang Hung[1,2]

1 Vietnam Maritime University, Hai Phong, Vietnam
2 Ural Federal University, Yekaterinburg, Russian

nkk@vimaru.edu.vn, hung.luuquang@vimaru.edu.vn

Abstract. The design of the directional coupler with reduced physical dimensions, using compact structures. The proposed coupler operates at two Central frequencies, 900 and 1900 MHz. The use of compact structures instead of quarter-wave segments made it possible to reduce the standard coupler by 67%, with minor performance degradation.

1. Introduction

In some cases it is necessary that directional couplers operate on several frequencies. This allows you to use such a device for several tasks at once. So the addition of idle loops in the standard design of the directional coupler allows to achieve the operation of the device in two operating frequency bands. Typically this frequency of 900 and 1800 MHz. At such frequencies, the traditional coupler takes up a lot of space, and for this reason, a procedure for miniaturization of its design is required. The literature describes many different techniques that can achieve good results in miniaturization, consider only a few works relating to the miniaturization of directional couplers. In [1] proposed to reduce the size using quasi-concentrated elements, in [2] equivalent transmission lines, in [3] U-shaped capacitances, in [4] periodic capacitive loads, asymmetric T-shaped structures in [5], low-pass filters in [6,7,8], slow-down systems in [9,10], artificial transmission lines in [11-13], fractal structures in [14,15], in [16] high-resistance elements, in [17] loaded loops, in [18-21] interdigital capacitors. In this paper, the design of the coupler was proposed by replacing four quarter-wave segments with compact structures having approximately similar characteristics at the Central frequency.

2. Design

Figure 1 shows the design of a directional coupler operating at one Central frequency, and figure 2 shows its frequency characteristics. The substrate material is a material with a dielectric constant of 4.4, a thickness of 1 mm and a loss of 0.02.

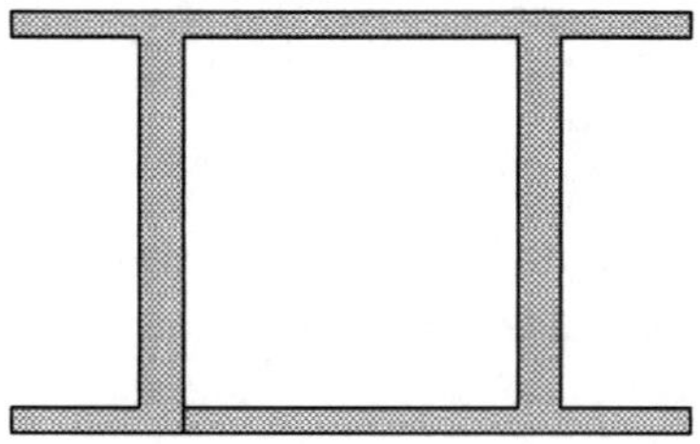

Figure 1. Topology directional coupler

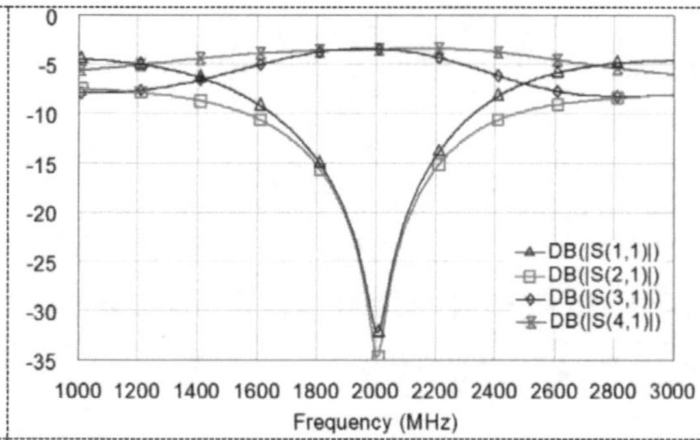

Figure 2. S-parameters from frequency

The addition of four idle loops, and the optimization of the design allows us to obtain the topology of the directional coupler [22], which divides the input power at two frequency ranges is 900 and 1800 MHz. The phase difference between the output signals at the two Central frequencies of the device must also be maintained. The topology of the coupler operating at two frequencies is shown in figure 3, and its frequency characteristics are shown in figure 4. After the standard design was designed and obtained, compact structures were synthesized, which allowed, after replacement, to reduce the overall dimensions of the coupler while maintaining its performance.

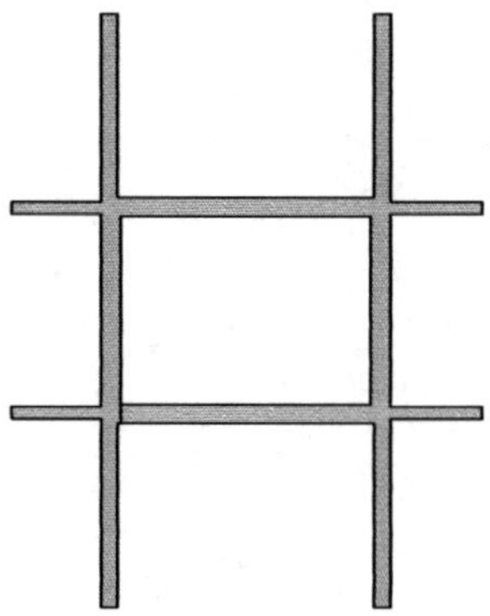

Figure 3. The topology of the dual-frequency directional coupler

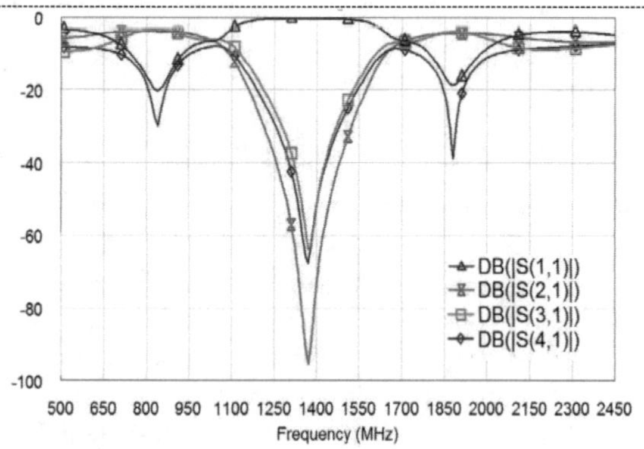

Figure 4. S-parameters from frequency

The area of a traditional coupler is 3020,3 mm^2. Operating frequency band estimated at the level of isolation -15 dB is 102 MHz. The second frequency band at the Central frequency of 1900 is 110 MHz. The imbalance between the output signals is 0.1 dB at 850 MHz and 0.15 dB at 1900 MHz. The compact structure is shown in figure 5 and its characteristics are shown in figures 6.7.

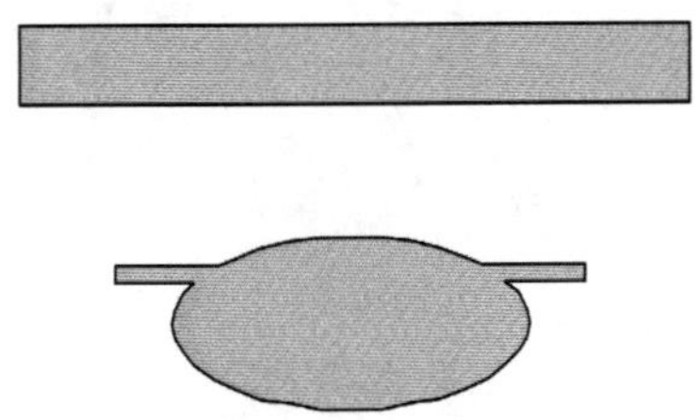

Figure 5. Topology of a quarter-wave segment and a compact structure with an impedance of 50 Ohm

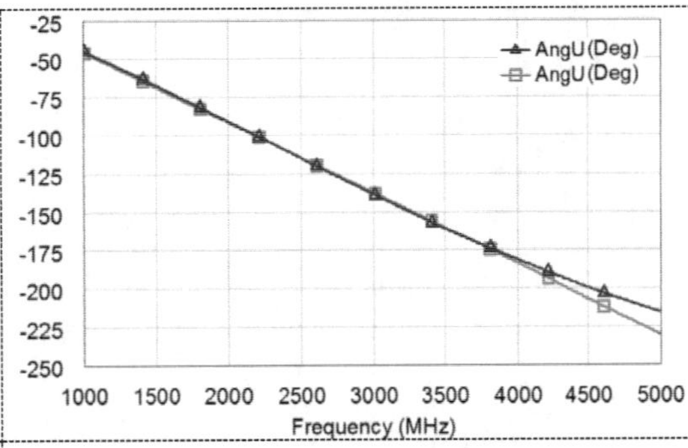

Figure 6. Phase on frequenc

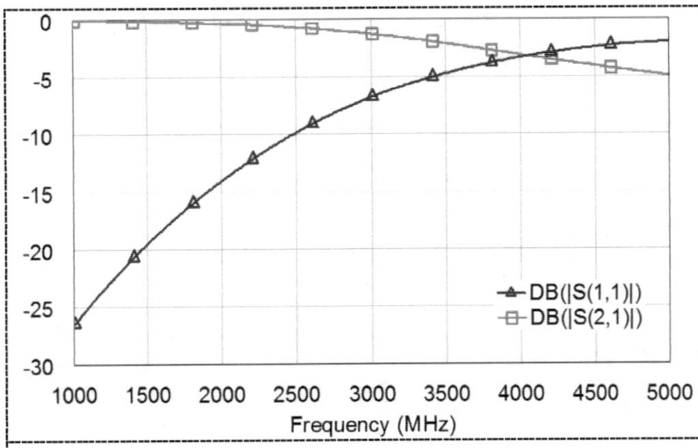

Figure 7. Frequency characteristics of a compact structure

Based on the frequency characteristics shown in figures 8.9, it can be seen that the phase in the required frequency band coincides up to 3500 MHz. The transmission coefficient of the structure begins to deteriorate significantly after the frequency of 2500 MHz. After calculating the necessary structures on a substrate with a permittivity of 4.4 mm and a thickness of 1 mm, the topology of a compact two-frequency coupler operating at frequencies of 800 and 1950 MHz was obtained. The frequency shift is caused because of the distinctive phase characteristic of compact structures.

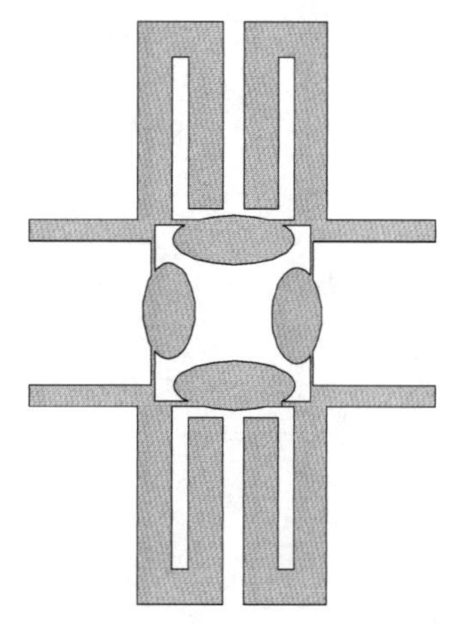

Figure 8. Compact dual-frequency directional coupler

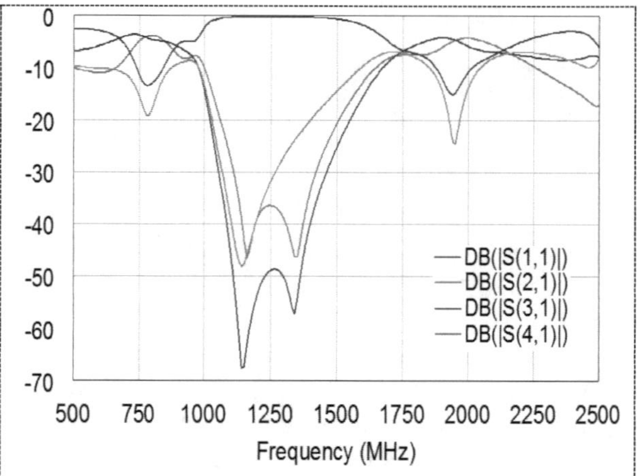

Figure 9. S-parameters from frequency

The area of the compact coupler is 998.6 mm^2. Operating frequency band estimated at the level of isolation -15 dB is 100 MHz. The imbalance between the output signals is 0.1 dB at 850 MHz and 0.15 dB at 1900 MHz. For comparison, the results were listed in table 1.

Table 1. Comparison of coupler

Design	Area, mm2	1 Bandwidth based on 15 dB isolation level, MHz	2 Bandwidth based on 15 dB isolation level, MHz
Standard	3020,3	102	110
Compact	998,6	75	90

3. Conclusion

A directional coupler with reduced dimensions and operating at two frequencies of 900 and 1900 MHz was designed and investigated. Due to minor differences in phase and amplitude at the required frequencies, the difference from the characteristics of the standard design is an increase in the imbalance, the displacement of the Central frequency. The area of the device was reduced by 67%.

References

[1] S.-S. Liao and J.-T. Peng, "Compact planar microstrip branch-line couplers using the quasi-lumped elements approach with nonsymmetrical and symmetrical T-shaped structure," *IEEE Trans. Microw. TheoryTech., vol. 54, pp. 3508-3514, Sep. 2006.*

[2] Letavin Denis, "Miniature microstrip branch line coupler with folded artificial transmission lines," *AEU - International Journal of Electronics and Communications, vol. 99, pp. 8-13, Feb. 2019. DOI: 10.1016/j.aeue.2018.11.016.*

[3] Letavin Denis A., "Compact branch-line coupler based on an inductor and a u-shaped capacitor," *19th International Conference of Young Specialists on Micro/Nanotechnologies and Electron Devices, EDM 2018, pp. 195-198, July 2018. DOI: 10.1109/EDM.2018.8435096.*

[4] K. W. Eccleston, S. H. M. Ong, "Compact planar microstripline branch-line and rat-race coupler," *IEEE Trans. Microw. Theory Tech., vol. 51, pp. 2119-2125, Oct. 2003.*

[5] S.-S. Liao, P.-T.Sun, N.-C.Chin and J.-T.Peng, "A novel compact-size branch-line coupler," *IEEE Microw.Wireless Compon.Lett. vol. 15, pp. 588-590, Sep. 2005.*

[6] Letavin Denis A., "Compact crossover based on low-pass filters," *19th International Conference of Young Specialists on Micro/Nanotechnologies and Electron Devices, EDM 2018, pp. 192-194, July 2018. DOI: 10.1109/EDM.2018.8434978.*

[7] Letavin Denis A., "Compact microstrip three-loop coupler," *2017 International Applied Computational Electromagnetics Society Symposium - Italy, ACES 2017, May. 2017, DOI: 10.23919/ROPACES.2017.7916036.*

[8] Letavin Denis A., "Miniature branch-line coupler structure analysis," *18th International Conference of Young Specialists on Micro/Nanotechnologies and Electron Devices, EDM 2017, July. 2017, pp. 99-101. DOI: 10.1109/EDM.2017.7981717.*

[9] J. Wang, B. Z. Wang, Y.-X.Guo, L.-C.Ong, and S. Xiao "A compact slow-wave microstrip branch-line coupler with high performance," *IEEE Microw.Wireless Compon.Lett. vol. 17, pp. 501-503, Jul. 2007.*

[10] Wei-Shin Chang and Chi-Yang Chang, "A High Slow-Wave Factor Microstrip Structure With Simple Design Formulas and Its Application to Microewave Circuit Design," *IEEE Trans. Microw. Theory Tech., vol. 60, no. 11, pp. 3376-3383, Nov. 2012.*

[11] Letavin Denis A., "Compact dual-frequency microstrip branch-line coupler using artificial transmission lines," *19th International Conference of Young Specialists on Micro/Nanotechnologies and Electron Devices, EDM 2018, pp. 185-188, July 2018. DOI: 10.1109/EDM.2018.8434988.*

[12] Letavin Denis A., "Two Methods for Miniaturization of Stub Quadrature Couplers," *Journal of Communications Technology and Electronics, vol. 63, Issue 8, Aug. 2018, pp. 933-935. DOI: 10.1134/S1064226918080107*

[13] Letavin Denis A., "Five-Stub Directional Device with Substantially Reduced Overall Dimensions," *2018 IEEE Radio and Antenna Days of the Indian Ocean, RADIO 2018. DOI: 10.23919/RADIO.2018.8572386*

[14] H.Ghali, T. A. Moselhy, "Miniaturized fractal rat-race, branch-line, and coupler-line hybrids," *IEEE Trans. Microw. Theory Tech., vol. 52, pp. 2513-2520, Nov. 2004.*

[15] J. Zhu, Y. Zhou, J. Liu, "Miniaturization Of Broadband 3-Db Branch-Line Coupler," *Progress In Electromagnetics Research Letters, vol. 24, pp. 169-176, 2011.*

[16] C. W. Tang, M.-G.Chen, "Synthesizing microstrip branch-line coupler with predetermined compact size and bandwidth," *IEEE Trans. Microw. Theory Tech., vol. 55, pp. 1926-1934, Sep. 2007.*

[17] Kimberley W. Eccleston, and Sebastian H.M. Ong, "Compact Planar Microstripline Branch-Line and Rat-Race Couplers," *IEEE Trans. Microw. Theory Tech., vol. 51, no. 10, pp. 2119-2125, Oct. 2003.*

[18] Kai-Yu Tsai, Hao-Shun Yang, Jau-Horng Chen, and Yi-Jan Emery Chen, "A miniaturized 2 dB Branch-Line Hybrid Coupler With Harmonics Suppression," *IEEE Microw. Wireless Compon. Lett., vol. 21, no. 10, pp. 537-539, Oct. 2011.*

[19] Ashmi Chakraborty Das, Lakhindar Murmu, Santanu Dwari "A Compact Branch-Line Coupler using Folded Microstrip Lines," *2013 International Conference on Microwave and Photonics (ICMAP).*

[20] Letavin Denis A., "Compact directional coupler with five stubs," *2018 IEEE Radio and Antenna Days of the Indian Ocean, RADIO 2018. DOI: 10.1088/1757-899X/524/1/012007*

[21] Kae-Oh Sun, Sung-Jin Ho, Chih-Chuan Yen, and Daniel van der Weide, "A Compact Branch-Line Coupler Using Discontinuous Microstrip Lines," *IEEE MICROWAVE AND WIRELESS COMPONENTS LETTERS, VOL. 15, NO. 8, AUGUST 2005*

[22] Kwok-Keung M. Cheng, and Fai-Leung Wong, "A novel approach to the design and implementation of dual-band compact planar 90 branch-line coupler," *IEEE Transactions on Microwave Theory and Techniques, vol. 52, no. 11, Nov. 2004.*

IOP Conf. Series: Materials Science and Engineering **630** (2019) 012017 doi:10.1088/1757-899X/630/1/012017

Small-sized Ring Coupler on Compact Structural Elements

Luu Quang Hung

Vietnam Maritime University, Hai Phong, Vietnam

hung.luuquang@vimaru.edu.vn

Abstract. In this work, the design of the ring coupler is investigated to reduce its size, with the possible preservation of its characteristics. So with the help of compact structural elements, the space occupied by the device was reduced, with a central operating frequency of 1000 MHz. The area of the device has been reduced by 80%, relative to the area of the standard device. With such a miniaturization, the frequency band was reduced by only 10%. The proposed miniaturization technique can be applied to any microstrip devices.

1. Introduction

Ring couplers, one of the types of couplers that can provide common-mode and anti-phase stimulation. Unlike two-loop directional couplers, the ring has a large area due to the fact that one of the segments has a length of wavelength. Therefore, this coupler is more critical to changes in the operating frequency. At low frequencies, the design can take up a lot of space, which is not always acceptable. Therefore, it is necessary to consider miniaturization, as a tool to eliminate this design flaw. To date, described a variety of different approaches, which, as a rule, consist in the replacement of standard segments with some structures that have smaller sizes, but similar characteristics. Were analyzed some of the work described in IEEE. In [1] proposed to reduce the size using quasi-concentrated elements, in [2] equivalent transmission lines, in [3] U-shaped capacitances, in [4] periodic capacitive loads, asymmetric T-shaped structures in [5], low-pass filters in [6,7,8], slow-down systems in [9,10], artificial transmission lines in [11-13], fractal structures in [14,15], in [16] high-resistance elements, in [17] loaded loops, in [18-21] interdigital capacitors. In this case, the possibility of miniaturization of an annular directional coupler using compact structural elements will be investigated.

2. Design

The first stage of the work was the design of a ring directional coupler on standard segments and an operating frequency of 1000 MHz. In this case, the substrate material was a dielectric with a dielectric constant of 4.4, a thickness of 1 mm and losses in the dielectric of 0.02. Figure 1 shows the topology of the ring coupler, with a Central frequency of 1 GHz. The frequency dependences for this design are shown in figure 2. The area of such a device is 3976 mm2, and the frequency band estimated at the level of 20 dB decoupling is 370 MHz.

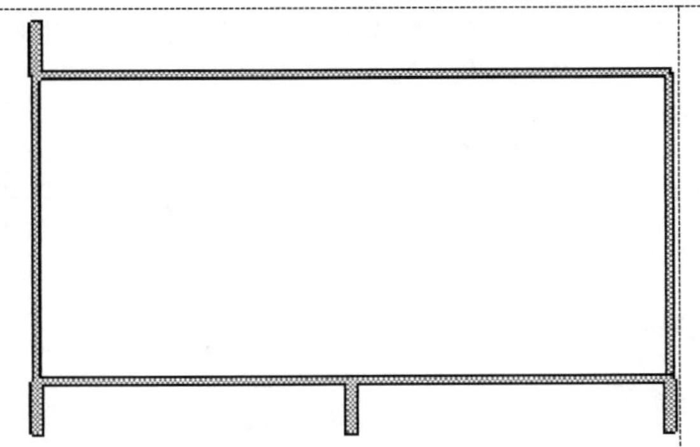

Figure 1. Topology directional coupler

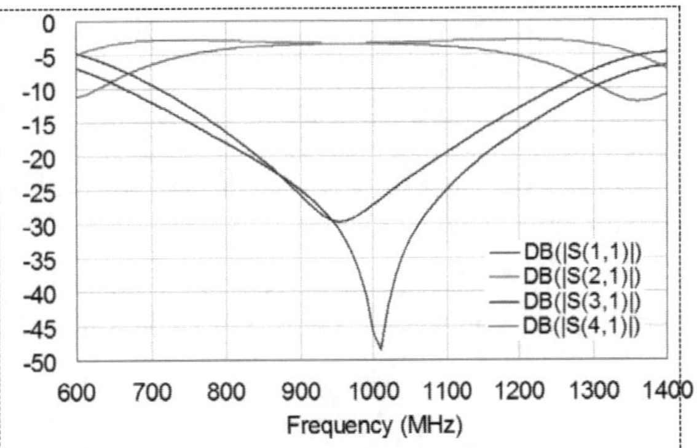

Figure 2. S-parameters from frequency

From the frequency characteristics it can be seen that the device operates at a Central frequency of 1000 MHz and has a frequency band of 370 MHz. The imbalance between the output signals does not exceed 0.05 dB. It should be noted that in its standard implementation it occupies a lot of space, so compact structural elements were synthesized, one of them is shown in figure 3. To assess the performance of such elements, their frequency characteristics were constructed, shown in figures 4,5.

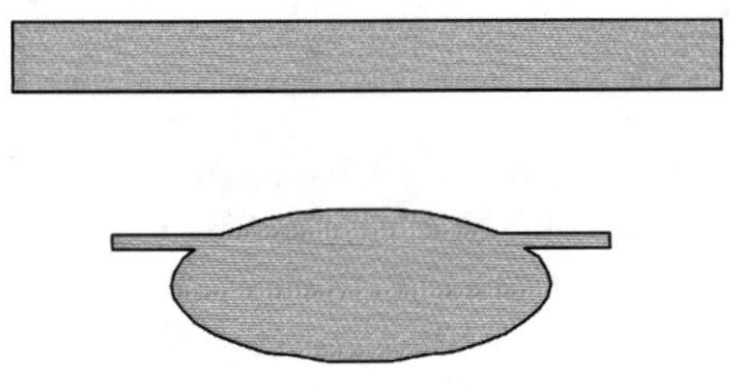

Figure 3. Topology of a quarter-wave segment and a compact structure with an impedance of 50 Ohm

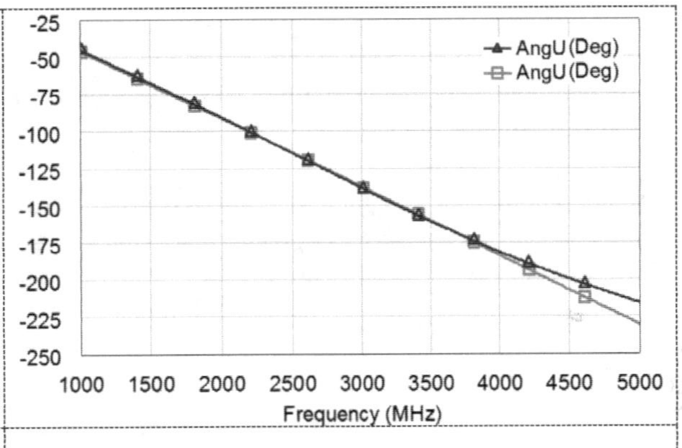

Figure 4. Phase on frequenc

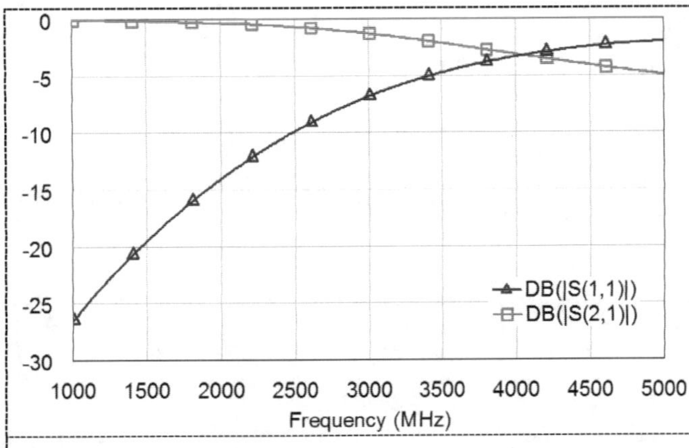

Figure 5. Frequency characteristics of a compact structure

IOP Conf. Series: Materials Science and Engineering **630** (2019) 012017 doi:10.1088/1757-899X/630/1/012017

These structural elements have identical characteristics at the Central frequency of the ring directional coupler. Therefore, each quarter-wave segment was gradually replaced by these elements. As a result, a compact topology of the ring directional coupler was obtained. Its topology is shown in figure 6. The area of such a device is 890 mm2, and the operating frequency band is equal to the value of 225 MHz. This is 40% less than the standard design. In this case, the area of the device is 77.6% less than the area of the standard design. The frequency characteristics obtained as a result of the calculation are shown in figure 7.

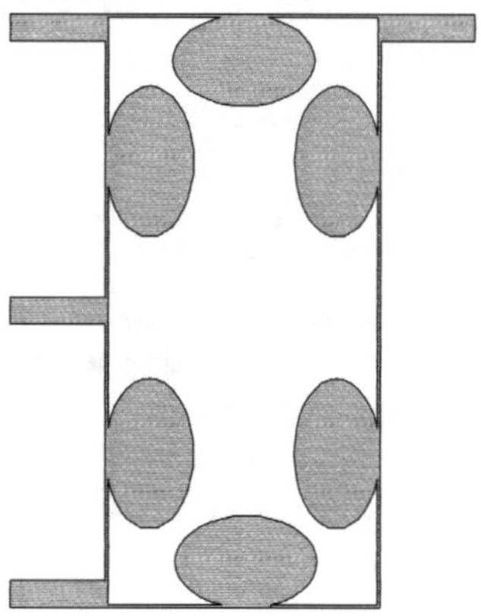

Figure 6. Compact coupler topology

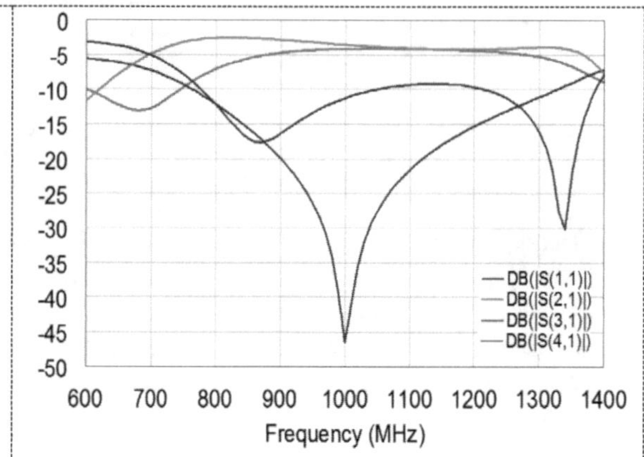

Figure 7. S-parameters from frequency

It can be seen that the imbalance between the output signals has increased, but the amplitude-phase dependences are at an acceptable level. It should also be noted that the internal space is used inefficiently for miniaturization, for this reason, high-resistance structural elements have been subjected to change, namely their bending while maintaining the same length. The modified topology of the compact ring directional coupler is shown in figure 8. The frequency characteristics obtained as a result of the calculation are shown in figure 9.

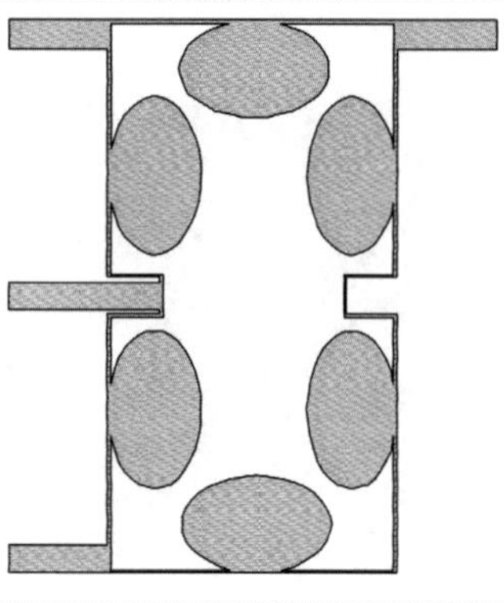

Figure 8. Compact coupler topology

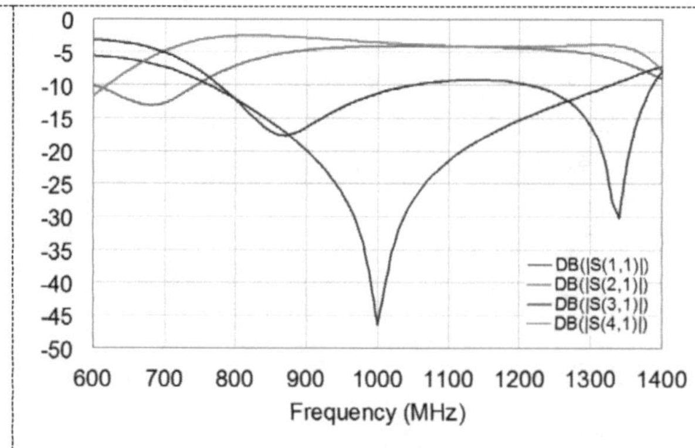

Figure 9. S-parameters from frequency

Compact coupler to the Central frequency of 1000 MHz. The area of such a device is 784 mm2, and the operating frequency band is equal to the value of 225 MHz. This area of the device is 80.3% less than the area of the standard design. The topology of the device is shown in figure 7, and its frequency characteristics are illustrated in figure 7.8. For comparison, the results were listed in table 1.

Table 1. Comparison of coupler

Design	Area, mm2	Bandwidth based on 20 dB isolation level, MHz
Standard coupler	3976	370
Compact coupler	784	225

3. Conclusion

In this work, a study was conducted aimed at miniaturization of the directional overwites ring, with a Central operating frequency of 1000 MHz. With designed compact structural elements, managed to achieve the miniaturization of the geometrical sizes of the tap with 784 mm2 to 3976 mm2. It turns out that the compact design is 80% less than the standard, but its operating frequency band decreased by 40%, and also increased the imbalance between the output signals in the device.

References

[1] S.-S. Liao and J.-T. Peng, "Compact planar microstrip branch-line couplers using the quasi-lumped elements approach with nonsymmetrical and symmetrical T-shaped structure," *IEEE Trans. Microw. TheoryTech., vol. 54, pp. 3508-3514, Sep. 2006.*

[2] Letavin Denis, "Miniature microstrip branch line coupler with folded artificial transmission lines," *AEU - International Journal of Electronics and Communications, vol. 99, pp. 8-13, Feb. 2019. DOI: 10.1016/j.aeue.2018.11.016.*

[3] Letavin Denis A., "Compact branch-line coupler based on an inductor and a u-shaped capacitor," *19th International Conference of Young Specialists on Micro/Nanotechnologies and Electron Devices, EDM 2018, pp. 195-198, July 2018. DOI: 10.1109/EDM.2018.8435096.*

[4] K. W. Eccleston, S. H. M. Ong, "Compact planar microstripline branch-line and rat-race coupler," *IEEE Trans. Microw. Theory Tech., vol. 51, pp. 2119-2125, Oct. 2003.*

[5] S.-S. Liao, P.-T.Sun, N.-C.Chin and J.-T.Peng, "A novel compact-size branch-line coupler," *IEEE Microw.Wireless Compon.Lett. vol. 15, pp. 588-590, Sep. 2005.*

[6] Letavin Denis A., "Compact crossover based on low-pass filters," *19th International Conference of Young Specialists on Micro/Nanotechnologies and Electron Devices, EDM 2018, pp. 192-194, July 2018. DOI: 10.1109/EDM.2018.8434978.*

[7] Letavin Denis A., "Compact microstrip three-loop coupler," *2017 International Applied Computational Electromagnetics Society Symposium - Italy, ACES 2017, May. 2017, DOI: 10.23919/ROPACES.2017.7916036.*

[8] Letavin Denis A., "Miniature branch-line coupler structure analysis," *18th International Conference of Young Specialists on Micro/Nanotechnologies and Electron Devices, EDM 2017, July. 2017, pp. 99-101. DOI: 10.1109/EDM.2017.7981717.*

[9] J. Wang, B. Z. Wang, Y.-X.Guo, L.-C.Ong, and S. Xiao "A compact slow-wave microstrip branch-line coupler with high performance," *IEEE Microw.Wireless Compon.Lett. vol. 17, pp. 501-503, Jul. 2007.*

[10] Wei-Shin Chang and Chi-Yang Chang, "A High Slow-Wave Factor Microstrip Structure With Simple Design Formulas and Its Application to Microewave Circuit Design," *IEEE Trans. Microw. Theory Tech., vol. 60, no. 11, pp. 3376-3383, Nov. 2012.*

[11] Letavin Denis A., "Compact dual-frequency microstrip branch-line coupler using artificial transmission lines," *19th International Conference of Young Specialists on Micro/Nanotechnologies and Electron Devices, EDM 2018, pp. 185-188, July 2018. DOI: 10.1109/EDM.2018.8434988.*

[12] Letavin Denis A., "Two Methods for Miniaturization of Stub Quadrature Couplers," *Journal of Communications Technology and Electronics, vol. 63, Issue 8, Aug. 2018, pp. 933-935. DOI: 10.1134/S1064226918080107*

[13] Letavin Denis A., "Five-Stub Directional Device with Substantially Reduced Overall Dimensions," *2018 IEEE Radio and Antenna Days of the Indian Ocean, RADIO 2018. DOI: 10.23919/RADIO.2018.8572386*

[14] H.Ghali, T. A. Moselhy, "Miniaturized fractal rat-race, branch-line, and coupler-line hybrids," *IEEE Trans. Microw. Theory Tech., vol. 52, pp. 2513-2520, Nov. 2004.*

[15] J. Zhu, Y. Zhou, J. Liu, "Miniaturization Of Broadband 3-Db Branch-Line Coupler," *Progress In Electromagnetics Research Letters, vol. 24, pp. 169-176, 2011.*

[16] C. W. Tang, M.-G.Chen, "Synthesizing microstrip branch-line coupler with predetermined compact size and bandwidth," *IEEE Trans. Microw. Theory Tech., vol. 55, pp. 1926-1934, Sep. 2007.*

[17] Kimberley W. Eccleston, and Sebastian H.M. Ong, "Compact Planar Microstripline Branch-Line and Rat-Race Couplers," *IEEE Trans. Microw. Theory Tech., vol. 51, no. 10, pp. 2119-2125, Oct. 2003.*

[18] Kai-Yu Tsai, Hao-Shun Yang, Jau-Horng Chen, and Yi-Jan Emery Chen, "A miniaturized 2 dB Branch-Line Hybrid Coupler With Harmonics Suppression," *IEEE Microw. Wireless Compon. Lett., vol. 21, no. 10, pp. 537-539, Oct. 2011.*

[19] Letavin Denis A., "Compact directional coupler with five stubs," *2018 IEEE Radio and Antenna Days of the Indian Ocean, RADIO 2018. DOI: 10.1088/1757-899X/524/1/012007*

[20] Ashmi Chakraborty Das, Lakhindar Murmu, Santanu Dwari "A Compact Branch-Line Coupler using Folded Microstrip Lines," *2013 International Conference on Microwave and Photonics (ICMAP).*

[21] Kae-Oh Sun, Sung-Jin Ho, Chih-Chuan Yen, and Daniel van der Weide, "A Compact Branch-Line Coupler Using Discontinuous Microstrip Lines," *IEEE MICROWAVE AND WIRELESS COMPONENTS LETTERS, VOL. 15, NO. 8, AUGUST 2005*

IOP Conf. Series: Materials Science and Engineering **630** (2019) 012018 doi:10.1088/1757-899X/630/1/012018

Bearing Fault Diagnosis Using Structure Optimized Deep Convolutional Neural Network under Noisy Environment

Guoqiang Jin, Dongjie Li, Ye Wei, Wenhui Hou, Huai'an Chen, Yi Jin and Chang'an Zhu

Department of Precision Machinery and Precision Instrument, University of Science and Technology of China, Hefei, Anhui, 230026, P.R.China.

jinyi08@ustc.edu.cn, jgq@mail.ustc.edu.cn

Abstract. In recent years, deep learning has been gradually applied in bearing fault diagnosis thanks to its powerful learning and representation ability. However, deep learning-based methods are usually computationally intensive and have redundant parameters. In this paper, we propose an end-to-end structure optimized deep convolutional neural network for bearing fault diagnosis using the local sparse structure. The width and depth of the network are increased and the redundant parameters are reduced while maintaining the representation capabilities and performance of the network. The local sparse structure increases the computational efficiency and reduces the risk of overfitting. The proposed method directly employed the raw signals as input without extra feature extraction with the ability that could distinguish both the bearing fault types and the corresponding severity. Experiments show that the proposed method achieves similar performance compared to the original method by using 46.47% parameters as used in the original study, even under the noisy environment.

1. Introduction

Rolling bearings are the critical parts in rotating machinery and their failure will lead to serious consequences [1]. Bearing fault diagnosis plays a key role in maintaining the normal operation of the machine. In the past few decades, bearing fault diagnosis based on data-driven methods have been widely carried out [2]. Early signal processing and learning-based methods have shown their weakness in dealing with signals in complex scenes. These methods require heavily handcrafted preprocessing procedures and expert knowledge. In recent years, deep learning has been gradually applied for fault diagnosis in bearings [3]. Deep learning has a good feature recognition and representation capability, and can automatically extract features without complex handcrafted preprocessing procedures.

The recent deep learning approaches in bearing fault diagnosis focus on extracting features from the raw signals without manual features extraction operation, providing an end-to-end classification method. In [4], the proposed adaptive overlapping convolutional neural network (AOCNN) took the augmented raw signals as the input and used the adaptive convolutional layer with overlapping and root-mean-square pooling layers to overcome the shift variant and marginal problems. Chen et al. [5] proposed a deep Inception net with atrous convolution (ACDIN) method that employed the Inception structure with the raw signal as input to classify the real bearing failure after training on the signals which were collected from the artificially damaged bearings. Zhang et al. [6] employed the convolutional neural network (CNN) as the feature extractor to mine features from the raw signal without complicated preprocessing. The CNN found to be a powerful feature exactor and classifier. In [7], the same author proposed the wide first-layer kernels convolutional neural network (WDCNN) to

IOP Conf. Series: Materials Science and Engineering **630** (2019) 012018 doi:10.1088/1757-899X/630/1/012018

classify the raw signal under noisy condition. However, these aforementioned methods are computationally intensive. They require a considerable large number of parameters. Inspired by the Inception structure [8], we presented a local sparse structure-based neural network for fault diagnosis in bearings which is computational efficiency and requires fewer parameters. Therein, the local sparse structure is used to replace the computationally intensive layers in the original deep neural network, especially the fully connected layer. The proposed method is an end-to-end neural network that took the raw signal as the input without handcrafted feature selection procedure and could classify the bearing fault types and the corresponding severity.

2. The basic theory of the neural network

The convolutional neural network [9] was inspired by the biological neural system that combines layers composed of multiple neurons. Each neuron in a layer denotes as a kernel that mainly focuses on a small reception area which greatly reduces the number of parameters compared to the fully connected network. The basic operation in CNN is the convolution operation. The CNN could extract features with variable length of the input and its computation cost is relatively low due to the weight sharing policy among the kernels in the same filter layer. The batch normalization (BN) is proposed to solve the internal covariate shift problems by normalizing the input distribution of a layer [10]. The BN layer is usually added after a convolutional layer before the activation function. The BN could accelerate the training process and improve the stability of the deep neural network.

The pooling uses the overall statistical characteristics of the adjacent regions to replace the output of the network in that region [11]. The pooling has the local invariant characteristic to the input and it reduces the parameters of the output. The maximum pooling and average pooling are two common pooling operations. The maximum pooling uses the maximum value in a local region to represent that region while the average pooling uses the average value of the local region.

3. The proposed structure optimized neural network

3.1. The Inception structure

The Inception structure was originally proposed to increase the representation capability, i.e. the depth and width of the network, while keeping the computational consumption. Directly increasing the depth of the neural network by stacking the layers would increase the network size dramatically and would lead to overfitting more likely. A reasonable way to solve the problem is to replace the computationally intensive layers by the sparse ones. Assuming the projection of some region in the input to a cluster of neurons in the later layers with high correlation, then we could use a layer of 1x1 convolutions to represent this cluster [12]. Similarly, if there would be a more spatially spread out clusters, then the clusters can be represented by a larger kernel in the convolutions, i.e. the 3x1 convolutions. The concatenation of these layers would have a similar feature representation as using the normal deep structure. One of the proposed local sparse structure is shown in figure 1(a).

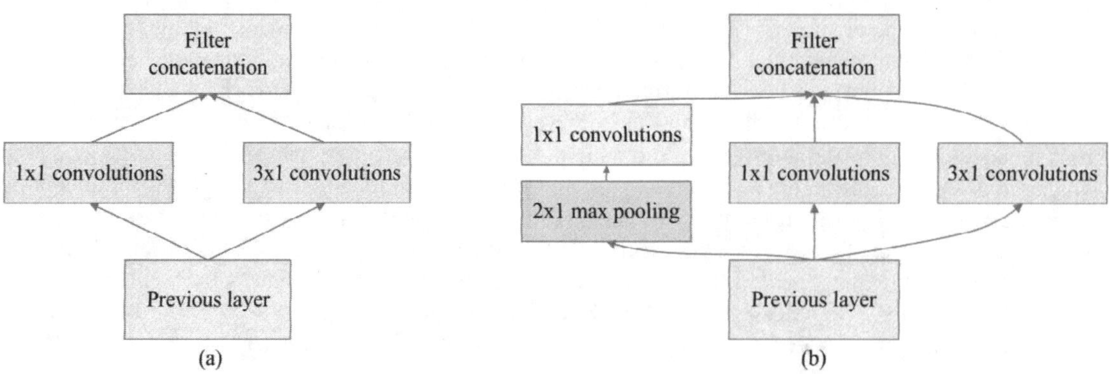

Figure 1. The proposed Inception module: (a) the naive version, (b) the dimensionality reduction version.

In order to improve the representation ability of network and to reduce the dimension of the concatenated layer, a parallel path for pooling operation followed by 1x1 convolutions is added. The 1x1 convolutions are employed to compress the high dimensional information, and are used as the nonlinear activation function after the pooling operation. The dimensionality reduction structure is depicted in figure 1(b). The stride of the 1x1 and 3x1 convolutions on the right is 2 in order to reduce feature map size. Specifically, the number of the filters in the independent 1x1 convolutions and 3x1 convolutions are 32 and 16 respectively for both the naive and dimensionality reduction versions. The number of filters in the 1x1 convolutions after the maximum pooling is 8. The Inception module improves network representation by increasing the width of the network while avoiding a dramatical increase in computational complexity.

3.2. The architecture of the proposed neural network

The CNN architecture has proven to be an effective method for the bearing fault diagnosis as reviewed above. However, due to large numbers of parameters to be calculated, the deep neural network based methods suffer the redundant calculation problem. We proposed a structure optimized network based on deep convolutional neural network using the local sparse structure to reduce the number of parameters and increase the computational efficiency while keeping the performance. The proposed architecture is shown in figure 2.

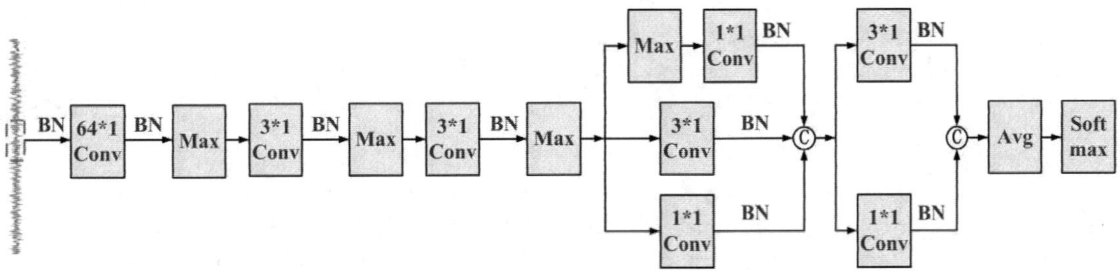

Figure 2. The architecture of the proposed neural network.

This proposed model is based on the WDCNN with the Inception structure improvement. Except for the last two local sparse structures and the average pooling operation in the network, the remaining parameters are the same as in the original network. The core idea of the proposed method is to replace the high dimensional convolutional layer and the fully connected layer in the deep neural network with the Inception structures as shown in figure 1. For convenience, we refer the structure in figure 1(a) as A, the structure in figure 1(b) as B, the structure before B shown in figure 2 is called the Base, and the convolutional block in the original WDCNN replaced by block B in figure 2 is called C. The proposed method is mainly composed of the Base block, the B block, and the A block. The effects of different combinations of these blocks are discussed in section 4.3. The Max and Avg modules in the figure represent the maximum pooling and average pooling, respectively. The © mark represents the concatenation operation to concatenate the features from different paths. Since there is no fully connected layer in the proposed network, the number of the network parameters has been reduced a lot compared to the original network, thus reduces the risk of overfitting.

4. Validation of the proposed neural network

4.1. Dataset description

The bearing fault dataset we used is from the Case Western Reserve University (CWRU) Bearing Data Center [13]. The vibration data is collected at the drive end with the sample rate of 48000Hz, and the motor load is 3HP. There are four types of health conditions: (1) normal condition; (2) ball fault; (3) inner race fault; (4) outer race fault. The fault sizes are 7 mils, 14 mils, and 21 mils (1 mil = 0.001 inches) for ball fault, inner race fault, and outer race fault types respectively. There are 10 different classes of data with 289 samples in each class, totally 2890 samples. We randomly select 80% samples as the training set and the rest as the test set. The input length of the network is 1670 which is calculated from the number of points collected by the motor with one rotating revolution. The amplitude of the raw signal is scaled to [-1, 1]. Noise is widespread, and the Gaussian white noise is one of the most representative and easily quantifiable noises. In order to evaluate the performance of the proposed neural network under noise condition, we add the additive white Gaussian noise with different levels of signal-to-noise ratio (SNR) into the raw data. The SNR is defined as follows:

$$SNR_{dB} = 10\log_{10}\left(\frac{P_{signal}}{P_{noise}}\right) \tag{1}$$

where P_{signal} is the power of the original signal and the P_{noise} is the power of the noise signal. We test the proposed method with the SNR of noisy signals ranging from −10 dB to 10 dB.

4.2. Training details

We compared the proposed network with the original WDCNN. Both methods used the same scaled raw signal as the input without further feature extraction procedure. The Adam [14] algorithm was used as the optimizer. The batch size was fixed to 64. We trained the original WDCNN and the proposed methods on signals without adding noise and test on signals with different levels of noise. Ten trials were conducted on each method to get the final result. The number of trainable parameters in the proposed method is 22356 while in the original WDCNN the number is 48112. In other words, our proposed method used 46.47% parameters as used in the original study. In order to study the effects of different structures on the network performance and computational cost as mentioned before, we also conducted experiments on different combinations of blocks and finally selected the best one.

4.3. Experiment result

The classification results for several different structures are shown in table 1. From the table, it can be seen that the A block has more parameters than the B block and is more suitable as the last block in the network. The proposed method (Base+B+A) has the best accuracy with small standard deviation and a relatively small amount of parameters compared the other combinations of structures, indicating that the proposed structure optimized network is efficient in reducing parameters while keeping the performance of the neural network.

Table 1. Comparison of accuracy with different structures at -4 dB SNR.

Structures	Original	Base+C+B	Base+C+A	Base+A+B	Base+B+A
Accuracy (%)	79.64 ± 4.70	77.44 ± 4.03	79.53 ± 3.10	78.81 ± 4.24	80.50 ± 2.91
Parameters	48112	28452	29196	22068	22356

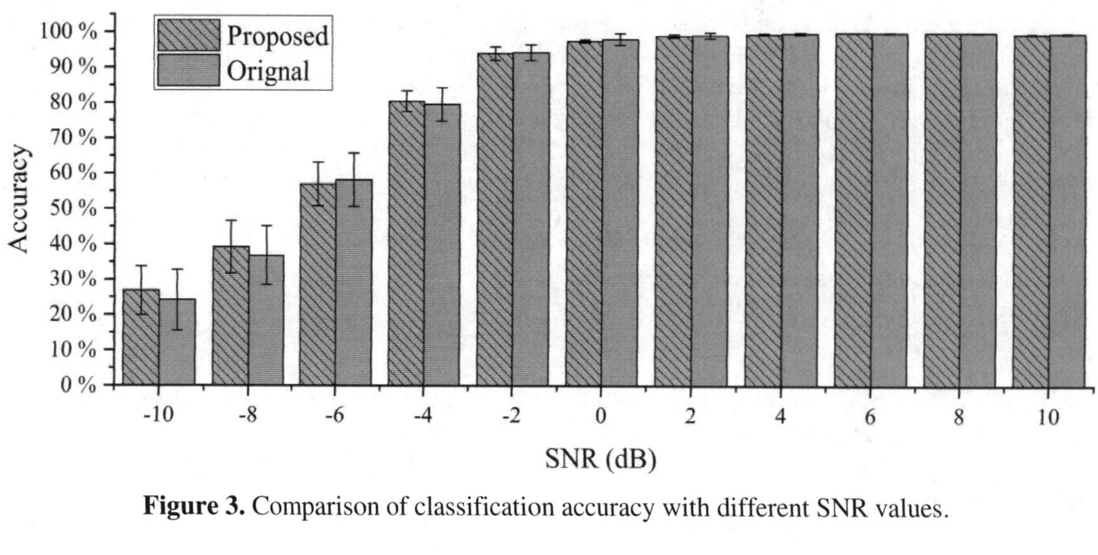

Figure 3. Comparison of classification accuracy with different SNR values.

The classification result is shown in figure 3. The error bar on each column indicates the standard deviation of each group of 10 trials. From the figure, it is clear that the classification accuracy drops and the standard deviation rises as the level of the noise increases. It is due to the reason that it is difficult to extract useful information from the noisy signals. The classification results indicate that the proposed method has almost the same performance as the original method, even slightly better in high noise conditions. The proposed structure optimized method reduces the number of parameters in the high-level layers of the network and reduces the risk of overfitting. After extracting features from low-level layers in the network, high-level layers networks with local sparse structure are easier to learn global information, thus improve the network representation.

5. Conclusion

In this paper, a structure optimized deep convolutional neural network with reduced parameters was proposed to diagnose the fault in rolling bearings. Experiments proved that by replacing the high dimensional convolutional layer and the fully connected layer with the local sparse structure can greatly reduce the number of network parameters without compromising the performance. The proposed deep convolutional neural network could distinguish not only the bearing fault types but also the corresponding severity. The proposed method is an end-to-end signal identification and classification neural network, which is a promising architecture to apply to other signal processing scenarios.

Acknowledgments

This work was supported by the National Natural Science Foundation of China (No. 51605464).

6. References

[1] Tandon N and Choudhury A 1999 A review of vibration and acoustic measurement methods for the detection of defects in rolling element bearings. *Tribol. Int.* **32** 469-480

[2] Cerrada M, Sánchez R, Li C, Pacheco F, Cabrera D, Valente de Oliveira J and Vásquez R E 2018 A review on data-driven fault severity assessment in rolling bearings. *Mech. Syst. Signal. Pr.* **99** 169–196

[3] Liu R, Yang B, Zio E and Chen X 2018 Artificial intelligence for fault diagnosis of rotating machinery: A review. *Mech. Syst. Signal. Pr.* **108** 33–47

[4] Qian W, Li S, Wang J, An Z and Jiang X 2018 An intelligent fault diagnosis framework for raw vibration signals: adaptive overlapping convolutional neural network. *Meas. Sci. Technol.* **29** 095009

[5] Chen Y, Peng G, Xie C, Zhang W, Li C and Liu S 2018 ACDIN: Bridging the gap between artificial and real bearing damages for bearing fault diagnosis. *Neurocomputing.* **294** 61–71

[6] Zhang W, Peng G and Li C 2016 *Advances in intelligent information hiding and multimedia signal processing* (New York: Springer International Publishing) pp 77–84

[7] Zhang W, Peng G, Li C, Chen Y and Zhang Z 2017 A new deep learning model for fault diagnosis with good anti-noise and domain adaptation ability on raw vibration signals. *Sensors.* **17** 425

[8] Szegedy C, Liu W, Jia Y, Sermanet P, Reed S, Anguelov D, Erhan D, Vanhoucke V and Rabinovich A 2015 *IEEE Conf. on computer vision and pattern recognition* (New Jersey: IEEE)

[9] LeCun Y, Bottou L, Bengio Y and Haffner P 1998 Gradient-based learning applied to document recognition. *P. IEEE.* **86** 2278–2324

[10] Ioffe S and Szegedy C 2015 *Proc. of the 32nd Int. Conf. on machine learning* vol 37, ed F Bach and D Blei (Lille: PMLR) pp 448–456

[11] Boureau, Y-Lan, Jean P and LeCun Y 2010 *Proc.of the 27th Int. Conf. on machine learning* (USA: Omnipress) pp 111–118

[12] Min L, Chen Q and Yan S 2014 Network in network *CoRR* **abs/1312.4400**

[13] Lou X and Loparo KA 2004 Bearing fault diagnosis based on wavelet transform and fuzzy inference. *Mech. Syst. Signal. Pr.* **18** 1077–95

[14] Kingma D P, and Ba J 2015 *3rd Int. Conf. for learning representations* (New York: Ithaca)

Natural Light Property-Based LED Lighting System to Maintain Human Circadian Rhythm

Geon-Woo Jeon[1], Seung-Taek Oh[2], Jae-Hyun Lim[1,*]

[1] Department of Computer Science & Engineering, Kongju National University, Cheonan-si, Chungcheongnam-do 32588, Korea
[2] Smart Natural Space Research Center, Kongju National University, Cheonan-si, Chungcheongnam-do 32588, Korea

[*]defacto@kongju.ac.kr

Abstract. The cyclic properties of natural light give an effect on human health, and the intensity and wavelength of natural light are known to be important factors to maintain human circadian rhythm. However, modern people who stay longer indoors are exposed to the fixed light environment of artificial lighting, resulting in circadian rhythm imbalance and subsequent sleep disturbances. In this paper, a LED lighting system that reproduces the natural light properties with day cycle is proposed to support the human circadian rhythm. To do this, the properties of natural light that keep changing were measured and analyzed every hour, and then the color temperature and wavelength ratio of natural light over time were reproduced by controlling each channel of LED lighting. In order to confirm the effect of the proposed lighting that supports maintaining a circadian rhythm, a confirmation experiment of melatonin change in the animal body was performed. The experimental results showed that the rats irradiated with the proposed lighting had an increased melatonin level by more than around 20ng /ml compared to the rats irradiated with the general lighting, thus confirming that the proposed lighting had a positive effect on sustaining the circadian rhythm.

1. Introduction

Natural light provides a constantly changing light environment according to season, time, and weather conditions [1]. Natural light has the characteristic of changing with a day cycle and it is closely related to human circadian rhythm. Circadian rhythm is a biological rhythm for adapting effectively to cope with changes in the external environment and plays an important role in the human being's day cycle of 24 hours. In addition, the wavelength and intensity of periodically changing natural light are known to be important factors for maintaining circadian rhythm [2]. These circadian rhythms are related to human health. When the circadian rhythm is disturbed, it affects the synthesis and secretion of melatonin, causing sleep disturbances and problems in the immune system. Therefore, it is necessary to preserve the habit of following the natural lights' cycle to maintain and to improve health [3]. However, modern people with long indoor activities are experiencing problems such as circadian rhythm imbalance as they are exposed to fixed light environment from artificial lighting for an extended time [4]. Previous studies to resolve such problem partially provided the color temperatures of natural light, but failed to reproduce the cyclic properties of natural light [5].

In this paper, a natural light-based LED lighting system that reproduces the daily cyclic properties of natural light is proposed to maintain the human circadian rhythm. The natural light that changes every hour was measured using a spectrophotometer and then color temperature and wavelength properties of natural light were analyzed and extracted over time. The color temperature and wavelength ratio properties in a day cycle were then provided to the indoor residents through LED lighting that could be controlled by each channel. The effect of the natural light-based LED lighting was confirmed through the experiment of melatonin changes in the animal body according to the application of the proposed lighting that supports the circadian rhythm.

2. Natural Light-reproducing LED Lighting System

2.1 Color Temperature and Short-wavelength Properties of Natural Light

The natural light was measured to check the properties of the natural light which are continuously changing at every moment and to extract color temperature and short-wavelength ratio by time. The properties of the natural lights were measured from sunrise (-60 min.) until sunset (+60 min.) at the rooftop of a 10-story building located at a latitude of 36.85 and longitude of 127.14. The spectrophotometer (CAS 140 CT, Instruments) that can measure spectral irradiance by wavelength (280-800nm) was used for the measurement. Figure 1 shows the measurement results of the natural light on the 1st of September 2017 (weather: clear). It represents short-wavelength (380-480nm) ratio, which is known to affect the sleep pattern of humans and color temperature of natural light, which changes at a day cycle. In addition, the color temperature and short-wavelength ratio of a general lighting are displayed together for comparison.

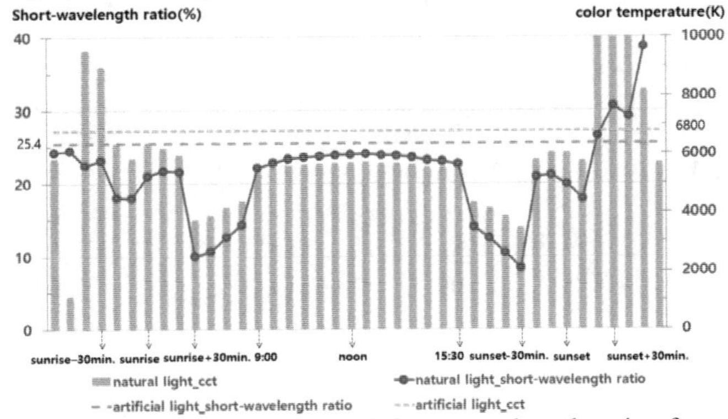

Figure 1. Properties of the color temperature and short-wavelength ratio of natural light and artificial lighting

In Figure 1, the short-wavelength (380-480nm) ratio was at a minimum of 8.4% at sunset (-30 min.), while it was a maximum of 23.9% at noon (12:00). Furthermore, the color temperature was at a minimum of 3485K at sunset (-30 min.), while it was at a maximum of 5717K at noon (12:00). However, the color temperature and the short-wavelength ratio showed an increasing tendency 30 min. before sunrise and 30 min. after sunset, but they were regarded as noises at the sunrise and sunset time zones and were excluded from natural light properties extraction. In addition, the H lighting from P company was measured and the results showed fixed light properties with the color temperature of 6800K and short-wavelength ratio of 25.4%.

2.2 Natural Light-based LED Lighting

In Figure 1 In order to reproduce the color temperature and short-wavelength ratio properties of the natural light over time, four-channel LED lighting which can be controlled by each channel was fabricated. In each channel, 500 LED light sources with a spec of 56 * 56 and with color temperatures of 3980K, 6800K, 2700K, and 5300K were applied, with the LED control module for each channel and wireless communication module for receiving lighting control packet implemented. The illumination was controlled by increasing the current by 16 from 0 to 255 per channel, with the total optical properties of 4,844 lighting control stages measured. The measurement results was confirmed that the optical property of a color temperature of 2700-6800K and short-wavelength ratio of 7.8-25.4% of the proposed lighting can be realized. The control indices of the lighting from 06:00 (AM) until 20:00 (PM) were extracted to reproduce the day cycle of natural light. For the LED control index, the control stage of the lighting nearest the natural light was explored by comparing the short-wavelength ratio and color temperature properties of natural light in each time zone with the optical property by each control stage. The control ratio by channel at the respective stage was selected after. Table 1 represents the reproduction results of the color temperature and short-wavelength of natural light measured by time through LED lighting.

Table 1. Example of lighting control by time zone

	Time	06:30	07:00	07:30	08:00	…	09:00 ~ 15:20	…	16:30	17:00	17:30	~20:00
Natural Light	Short wavelength ratio	9.9%	14.1%	17.1%	19.4%	…	22.6%	…	16%	12.8%	8.8%	-
	color temperature	3692K	4324K	4726K	5039K	…	5500K	…	4548K	4126K	3473K	-
Proposed Lighting	Short wavelength ratio	9.9%	14.1%	17.1%	19.4%	…	22.6%	…	16%	12.8%	8.8%	-
	color temperature	2940K	3568K	4126K	4805K	…	5674K	…	4139K	3461K	2814K	2814K
	control stage	(0, 128, 16, 64)	(32, 96, 48, 32)	(16, 80, 80, 16)	(80, 32, 48, 48)	…	(32, 32, 112, 16)	…	(16, 80, 80, 16)	(80, 32, 48, 48)	(96, 16, 48, 32)	(128, 0, 32, 0)
	Control result	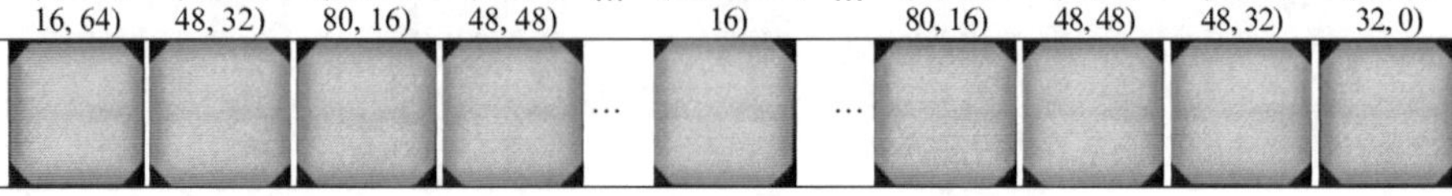				…		…				

LED lighting control was realized in the 06:30-09:00 and 15:30-17:30 zones to make short-wavelength ratio match the color temperature closest to the natural light based on the actual measurement data of the natural light. However, the 06:00-06:30 zone is where the intensity of illumination is prominently low, thus the lighting was made off. Meanwhile, the 09:10-15:20 zone is where the changes of optical property are meager; therefore, the average optical property of the respective zone was reflected. In addition, after 17:30, the control stage where the short-wavelength ratio was the lowest was implemented to aid in sleep.

3. Experiments and Analysis

An animal experiment was conducted to analyze the changes in melatonin in the animal body in order to confirm the effect of the proposed lighting which supports the human circadian rhythm. 20 rats were used for the experiments. During a week's acclimatization, the lighting which reproduced the property of natural light was irradiated to the experimental animals. The rats were grouped into ten each for the experimental group and control group. Two dark cases were fabricated to block external light. The proposed lighting and general lighting were installed at the top of the dark cases. The experiment proceeded by rearing the experimental rats under the environment with the proposed lighting and the control rats under the natural lighting. The rats were irradiated with each lighting daily from 06:00 until 20:00 for two weeks' experimental period. The changes in melatonin were measured by taking blood from the rats nine times at 23:00, 02:00, and 06:00 on the experiment starting date, after seven days, and on the completion date. The melatonin changes are shown in Figure 2.

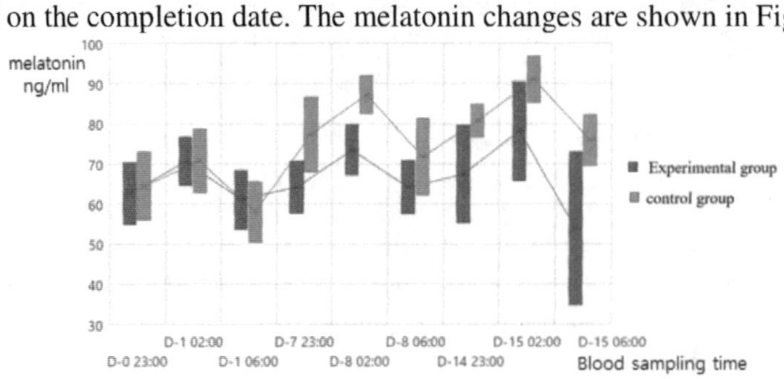

Figure 2. Changes in melatonin between the experimental group and control group

For analysis of the experimental results, Melatonin's assay kit was Abcam (ab213978). Statistical analysis was performed using Prism 5.03 (GraphPad Software Inc., San Diego, Calif., USA). The melatonin changes between the experimental group and the control group on the experiment starting date (D-0) were not significant. However, there were significant differences in the melatonin changes between the experimental group and control group on the 7th day (D-7) and 14th day (D-14) of the experiment. In the experimental group where the proposed lighting was implemented, the melatonin was increased by an average of 20ng/ml compared with that on the experiment starting date. In the case of the control where the general lighting was implemented, however, there were no significant differences in the melatonin amount.

4. Conclusion and Future Study Direction

In this paper, a LED lighting system that reproduces the properties of natural light was proposed to maintain the circadian rhythm of humans. The properties of natural light were measured and analyzed using a spectrometer whereas the color temperature and wavelength ratio of the natural light that keeps on changing daily were extracted. In addition, a four-channel LED lighting that can be controlled by channel was fabricated to reproduce the properties of natural light. After that, the color temperature and short-wavelength of the natural light were realized through LED lighting. In the 06:30-9:00 and 15:30-17:30 time zones, a LED lighting that matched the short-wavelength and was nearest to the color temperature of natural light was realized. Meanwhile, in the 09:10-15:20 time zones where the changes in the properties of natural light were meager, the average light properties in the respective time zones were reproduced. Furthermore, in the zone after sunset where properties of the natural light did not exist, the LED lighting which has the minimum short-wavelength ratio was provided to aid in sleep. An animal experiment was also conducted in the environment where the proposed lighting and general lighting was categorized to confirm the effect of the proposed lighting that supports the human circadian rhythm. The experimental results showed that in the experimental group where the proposed lighting was applied, the melatonin amount that is closely related to the circadian rhythm was increased by more than an average of 20ng/ml compared with that of the control group. It was, therefore, confirmed that the proposed lighting can help in maintaining the human circadian rhythm.

It is essential to conduct research in the future to develop a natural lighting reproducing a lighting system that can correctly reproduce the short-wavelength ratio and color temperature of natural light in real-time and can realize the changes of natural light by time.

Acknowledgments

This work was supported by the National Research Foundation of Korea (NRF) grant funded by the Korea government (MSIP) (No. 2017R1A2B2005601). The research was supported by the International Science and Business Belt Program through the Ministry of Science and ICT (2015-DD-RD-0068-04).

References

[1] Jeong Hak-Geun, Jang Cheol-Yong. A study on Color Rendering Improvement of the LED Lighting using Sun light. *Korea Solar Energy Society Conference Proceedings*, (2010.11): 91-95.

[2] George C. Brainard et al. Sensitivity of the human circadian system to short-wavelength(420-nm) Light. *Biological Rhythems*23(5), (2008): 397-386.

[3] Lee Hyun-Ah, Cho Chul-Hyun and Kim Leen, Human Circadian Rhythms. *Sleep Medicine and Psychophysiology*, 21.2(2014): 51-60.

[4] Kim Kyoung-Sil, Choi An-Seop. Study of Relationship between Light and Melatonin. *Proceedings of KIIEE Annual Conference*, (2011.11): 7-8.

[5] Moon Seung-Mi et al. Implementation of Smartphone-based Color Temperature and Wavelength Control LED Lighting System. *Cluster Computing*, 19.2(2016):949-996.

The modeling of the efficiency in the new generation manufacturing-distributive systems based on the cognitive production factors

Irina Omelchenko, Pavel Drogovoz, *Evgeniya Gorlacheva*, Vladimir Shiboldenkov and Olga Yusufova

Bauman Moscow State Technical University

gorlacheva@yandex.ru;

Abstract. The paper envisages the analysis of the new generation manufacturing-distributive systems. The particular attention is paid to the production function as the analytic instrument allowed evaluating the interrelation between economic results of an enterprise and production factors, which includes the cognitive production factors as well. The rationale for the enlargement of the traditional production functions by means of the transfer from the multiplicative to the logistic dependence has been analyzing. The results of production function modelling on the base of the logistic dependence have been depicted on the open data of a high tech enterprise from 2009 till 2018. The results of production function modelling have shown that the usage of logistic dependence has allowed tracing more precise the transition of production factors to the other path dependency. The inclusion of the cognitive production factors has allowed evaluating their contribution to the economic results of a high-tech enterprise. This work is still in progress and the presented results are preliminary and would be added and specified by the additional research.

1. Introduction

The existing challenges of the modern industrial manufacturing deal with the provision of the economic efficiency of mass production in transition to the agile customizable manufacturing-distributive systems. That can be called the systems of the new generation [1]. The relevance of its development is explained by the vast expansion of IT-technologies that determine the economic efficiency level of these systems as economic entities. In particular it is distinctly traced by the activity of high-technology enterprises that can be considered as the prototype of such systems.

High-tech enterprises are characterized by the increasing division of “material” and “non-material (intellectual)” manufacturing components; the establishment of intellectual part as a standalone product; the reinforcement of the globalization in the process of a new product [2]. Attended in these circumstances technical and economic barriers, imbalances in innovation-driven development connect with the inefficiency of existing production factors and the appearance of new ones, among them cognitive production factors as well. Under the cognitive production factors there is an amount of knowledge, capabilities that provide the gain in productivity while manufacturing product commercially successful [3].

Traditional approaches to the evaluation of the production factors efficiency from one side allow taking into consideration all effects in one indicator, from the other side assumptions that lay in the base of economic and mathematical models are often open to the fair criticism, as they don't consider the incremental and cyclic dynamics of the manufacturing-distributive systems of the new generation [4].

The processes of reciprocal transitions in qualitative and quantitative changes in the development of complex dynamic systems are modelled by the logistic curves [5]. Due to the interchange of evolutionary and revolutionary processes their dynamics presents the incremental and cyclic process. Thus there is the necessity of the modified production function elaboration and the modelling of the efficiency of the new generation manufacturing-distributive systems on the base of open data of a high-tech enterprise.

The paper has the following structure: part 1 presents the review of main approaches to the usage of the production function and depicts the peculiar characteristics of logistic curve; it also describes the notion of cognitive production factors and its typology; part 3 deals with the description of high-tech enterprises and peculiarities of Russian high-tech sector; part 4 presents the modelling procedure on the open data; part 5 finishes the paper with the results and conclusion.

2. Theoretical Background

2.1. Main approaches in the production function modelling

In the economic modelling there are two basic approaches in the evaluation of manufacturing efficiency: the total factor productivity (TFP), so called R. Solow's residual and the production possibility frontier (PPF) [6-9].

In Solow's approach it is assumed that the technological progress is exogenous and is determined as the part of output growth, not explained by the accumulation of production factors [4, 9]. This component, Solow's residual is the estimation of the technology progress. In Solow's approach the basic assumptions are following: the constant returns to scale in economy, the functioning of enterprises in terms of the perfect competition, the functioning of economies on the frontier of manufacturing capabilities. These assumptions are wide open to criticism [10, 11]. In real conditions one can register not only constant but increasing returns to scale. This issue has been considered in framework of so called PPF's approach that presents the non-parametric envelope curves, based on linear programming methods and the stochastic frontier of manufacturing capabilities intending the usage of panel data.

In the method of non-parametric envelope curves the efficiency level is measured by the degree of the potential implementation, i.e. the remoteness of a high-tech enterprise from its frontier of manufacturing capabilities. This method constitutes in the evaluation of the output deviation functions and also in the search of the maximum possible output. The main limitation of this approach is the deterministic character. This approach doesn't assume the "noise" in data; it is considered that all combinations "factors-output" presented in the observation can be achieved.

The approach of PPF described in [4, 9] allows taking into consideration the presence of random effects including measurement errors, that influence the deviation from the frontier. The influence of these random effects is taken into account by the decomposition of model residuals into the "noise" and "inefficiency" component, so called the extent of output deviation from its frontier level, which could be achieved in the given volume of production factors. The method of the stochastic frontier supposes that enterprises act not on the frontier but inside their manufacturing capabilities.

This method allows reflecting two aspects: the shift of manufacturing capabilities' frontier (scientific and technical progress) and the movement of an manufacturing-distributive system along this frontier (enterprise's efficiency dynamics). The inefficiency arises when an manufacturing-distributive system is inside of manufacturing capabilities' frontier and the output can be increased with the existing technologies and production factors. The farer the high-tech manufacturing-distributive system is from manufacturing capabilities' frontier the higher its inefficiency.

The conducted review has shown that these approaches have been used for evaluating aggregative manufacturing-distributive systems' efficiency on the country level. However in these approaches there isn't been taken into account the incremental and cyclic dynamics of the scientific progress on the level of a high-tech manufacturing-distributive system. Meanwhile mainly high-tech manufacturing-distributive systems are drivers for any national economy.

2.2. *The peculiarities of logistic curves*

The usage of the logistic curve one could find in works of many researches [5, 12, 13]. The logistic curve possesses the following peculiarities:

- The growth of investigated indicator with the growth of the argument;
- Changes in growth rate (increase and decrement);
- The growth ceiling.

The complexity of taking investing decisions deals with the fact that the choice of this or that set of production factors will determine the effect of its implementation – the ratio that determine the lifecycle phase of cognitive production factors in the moment of investment. As in a every time a high-tech manufacturing-distributive system has a limited set of cognitive production factors that could be supported by investment resources, thus the investment process is characterized by the discrete values of results putting at odds short-term and long-term investment goals. One of the problems is the forecast of the growth ceiling some of competitive cognitive production factors, designed for solving the same economic task.

The research of cognitive production factors asks the question whether the law of decreasing returns is correct. In short-term this law is senseless, as the increase of cognitive production factors supposes the modernization of manufacturing and the increase of employees' qualification, i.e. the changes of all production factors. Alongside with this the law of decreasing returns relate to the long-term period is trustful at the definite stages of the life cycle.

Taking into account that the second order derivative of the generalized logistic curve changes sign several times, the increasing and decreasing returns also change each other by turns [5,12,13]. This circumstance coincides with the obsolescence of production factors and cognitive ones as well. The usage of the modified production function on the base of the logistic curve allows evaluating the efficiency of high-tech manufacturing-distributive systems and also forecast the tipping points in the increasing and decreasing returns.

2.3. *Cognitive production factors*

In terms of modern economy the innovations consider as one of main sources for economic growth. Dynamic innovation development of high-tech manufacturing-distributive systems leads to the significant changes in national economy structure and determine the priorities of scientific and technological development of a country [2, 14-17].

The Russian high-tech manufacturing-distributive systems are the object of research for many Russian scientific schools [2, 15, 17, 18]. High-tech manufacturing-distributive systems are researched and analyzed from the perspective of its contribution to gross domestic product; amount of employees in high-technology sector; a share of R&D expenses in total expenses of a high-technology manufacturing-distributive system; a share of scientific extensive product in export flows.

However, namely the micro level of high-technology sector, the economy of a separate manufacturing-distributive system determines the development of high-technology sector as a whole. Its insufficient efficiency is due to the imbalance of innovation development, imbalance in productivity, imbalance in the usage of production factors.

According to [12] one additional percent of labour expenses increases output three time more than one additional present of capital expenses that impose the necessity to use inner resources in the achievement of strategic goals of a high-tech manufacturing-distributive systems.

High-tech manufacturing-distributive systems facing strategic tasks increase the relevance and practical meaning of production factors research, the nature of its impact to the results of a high-tech manufacturing-distributive system as an economic entity. The shaping factors are knowledge, know-how, R&D, i.e. intellectual component of manufacturing capabilities.

The premises of production factor theory are that the cost-effectiveness balance determines the strategic position of a manufacturing-distributive system in the market space. Conceptual issues of the modern production factors theory are following:

- Economic sustainability of a high-tech manufacturing-distributive system determines by the cost-effective balance of existing factors and their effective management;
- Competitive advantages depend on the availability of strategic resources;
- Effective management of existing production factors is provided by the organization possibilities of a high-tech manufacturing-distributive system.

The question has been arisen, what peculiarities should have production factors that the innovation-driven development would run effectively and intensively?

The cyber economy has changed dramatically the traditional production factors. If in industrial economy the shaping factors were the labour and the capital, in the post-industrial economy the shaping factors are high technologies and high rates of technological renewal. As the consequence the economic efficiency depends on the quality of human capital and a developed IT-infrastructure.

The amount of specialized knowledge, skills, and competences to transfer them in a competitive product allows singling out new type of production factors – cognitive production factors.

In scientific literature [19-24] the production factors are usually classified as basic, key, leading, strategic, shaping, etc. in this paper we use the typology of cognitive production factors suggested in [1].

The main criterion of distinguishing is the dichotomy "material/non-material" and "active/intentional". These two measures form four quadrants of the indexer, each from which coincides with the definite type of cognitive production factors (fig.1).

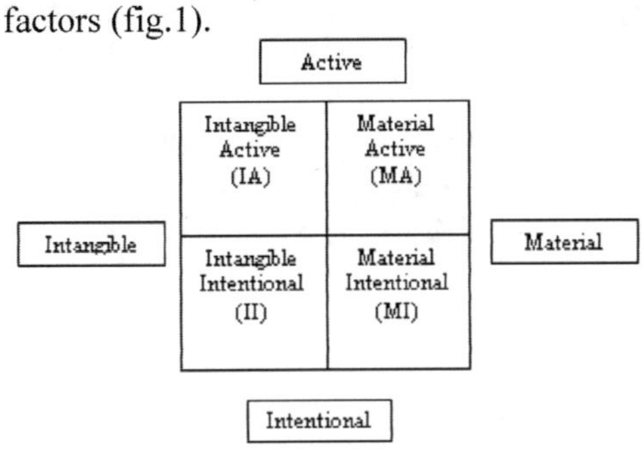

Figure 1. Types of cognitive production factors.

To material-active one can relate those cognitive production factors that have a material form and use in the economic activity of a high-tech manufacturing-distributive system, such as local computer networks, flexible manufacturing systems, robots, automated storage systems, planning systems, design systems, electronic document flow systems, systems of technical vision, etc.

To non-material-active one can relate intellectual property issues: know-how, licenses, data base, relations with clients, suppliers, etc.

To material-intentional one can relate potential advanced technologies, such as augmented reality technologies, artificial intelligence technologies, fog computing, deep learning, etc.

To non-material intentional one can relate experience, education level of employees, digital literacy, capacities for analysis, reflection, self-regulation, communicative skills, etc.

This list of cognitive production factors is not comprehensive and could be added and enlarged.

3. Characteristics of high-tech manufacturing-distributive systems

According to the Strategy of Economic Development in the Russian Federation (www.economy.gov.ru) [25], innovative economy is of high priority for Russia's long-term strategic development. In this regard, high-tech manufacturing-distributive systems play an important role. Below we will present a brief overview of the Russian high-tech sector based on existing studies [26, 27] and official documents [28-30].

IOP Conf. Series: Materials Science and Engineering **630** (2019) 012020 doi:10.1088/1757-899X/630/1/012020

The share of the Russian high-tech sector in world trade accounts for about 1,3% [31,32]. The most developed industries are defense, space, and aviation, which can be explained by the rich industrial heritage and a significant scientific potential. The share of the high-tech sector in the gross domestic product is quite large and constitutes 21% (see fig. 2) [30].

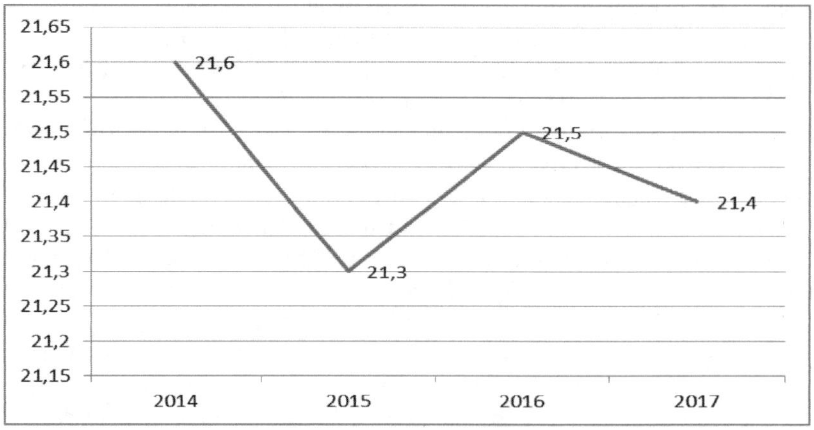

Figure 2. The share of high-tech products in Russia's GDP.

The structure of Russian high-tech export is rather heterogeneous. The largest share in the output of the high-tech sector is composed of commercial products of the aviation and space industries, as well as the products of defense industry. The proportion of products of the electronic complex is 14%, the nuclear industry - 13.6%.

Thus, the export of Russian high-tech manufacturing-distributive systems is concentrated within the narrow specialized areas of high-tech products, while the competitiveness of these manufacturing-distributive systems is several times lower in commodity groups designed for a mass end user.

Following [33, 34], the Russian high-tech sector is structured by the defence industry (28,6%), the space and avia industries (21,2%), the information and communication industry (18,9%), the medical equipment industry (14,3%), etc (see fig. 3).

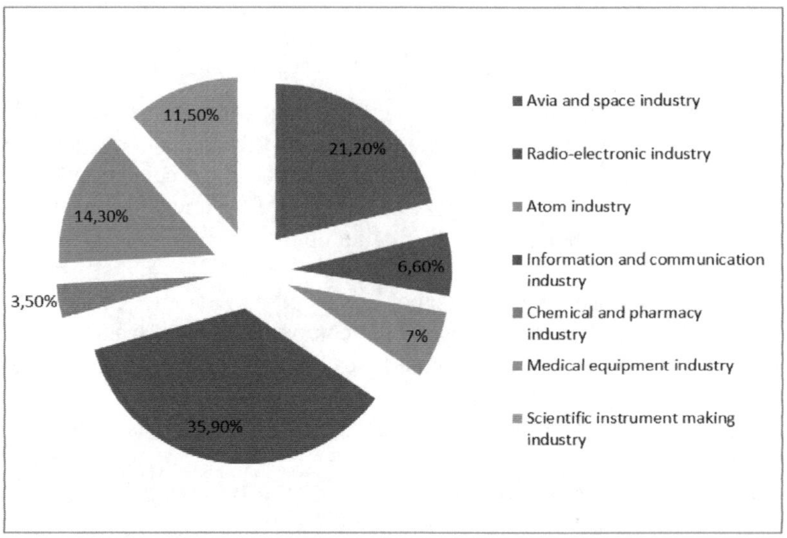

Figure 3. Structure of the Russian high-tech sector.

The distinctive characteristic of the Russian high-tech sector is the numerous industrial chains that are actively developing at the moment. At the same time, there has appeared a large number of ideas supported by different kinds of multidisciplinary R&D groups, although their significant efforts do not always lead to the commercialization of these ideas. However, in comparison with the manufacturing-distributive systems of medium technological industries, high-tech manufacturing-distributive systems lose in terms of the share of innovative products in the total output. This happens due to significant investment and innovation lags in the high-tech sector and the low performance of the Russian high-tech sector.

In the Russian Federation, the key areas of scientific and technical progress are reflected in the Priority Industries List [33] for the development of science and technology that generally comply with the world scientific and technological priorities: information and telecommunication systems; life sciences; the industry of nanosystems; transport and space systems; resource saving technologies; energy efficiency, energy saving nuclear energy. The development of these areas is aimed at shaping the non-defense high tech sector in the Russian economy.

Summing up, we could formulate the following characteristics of the high-tech sector in Russia:

- The high-tech sector influences the general economic development in Russia;
- The defence, aviation, and space industries give the largest input in the Russian high-tech export;
- Most high-tech manufacturing-distributive systems are still in state ownership as any efforts in attracting private investments have been hindered by the nature of the high-tech sector;
- Multidisciplinary R&D groups generate a great number of ideas, although commercialization does not always take place due to considerable innovation lags.

4. An Efficiency model of a high-tech manufacturing-distributive system on the base of the modified logistic curve

Taking into account that the main goal of a high-tech manufacturing-distributive system is the output of innovation product its performance function is the profit maximization of its output. The factors of the production function are capital, labour and cognitive production factors. The dependence is based on the logistic curve; the level of saturation is determined by the development level of production factors. The growth of factors is determined by the investments with the distributed lag. The suggested model is the description of the replenishment cycle, taking into account the incremental and cyclic dynamics of a high-tech manufacturing-distributive system:

$$Yt = S(K_t, L_t, CPF_t) \quad (1)$$

$$S(K_t, L_t, CPF_t) = Ke(-a(K_t, L_t, CPF_t)/1+K(e(-a(K_t, L_t, CPF_t)) \quad (2)$$

Considering the fact that the cognitive production factors are qualitative notion there is a necessity to make them quantitative. Table 1 presents their correlation with the quantitative economic parameters.

Table 1. The correlation of the production factors and the economic parameters.

Production factor	Economic parameter	Source of information
Capital	Own long-term capital, own circulating capital	Balance sheet, form1
Labour	Labour expenses with social tax remission	Balance sheet, Form 4
Cognitive production factors	Residual value of basic assets (MA), a share of property funds in R&D (MI); intellectual property value (IA); turnover rate of R&D personnel (II)	Balance sheet, Form 1, Form 2

In order to check the correctness of the logistic dependence between production factors and the results of a high-tech manufacturing-distributive system we have taken open data of one high-tech enterprise [35] for a period from 2009 till 2018. These prepared data for modelling are presented in table 2.

Table 2. Open data of a high-tech manufacturing-distributive system

Year	Capital	Labour	Cognitive production factors
2009	77.8	2	89.84
2010	66.5	2.4	83.8
2011	71.3	5.6	84.43
2012	72.77	15.42	85.71
2013	68.96	16.74	83.88
2014	73.84	16.56	86.27
2015	86.97	14.53	93.91
2016	106.23	16.13	106.5
2017	104.3	17.6	113.81
2018	126.1	19.15	129.65

4.1. Results of modeling

In the process of modelling we have singled out two stages: the logistic primary curve without the time period shift and with the time period shift and the logistic secondary curve without the time period shift and with the time period shift. The results of the first phase are presented on fig.4,5. The results of the second phase are described on fig.6,7.

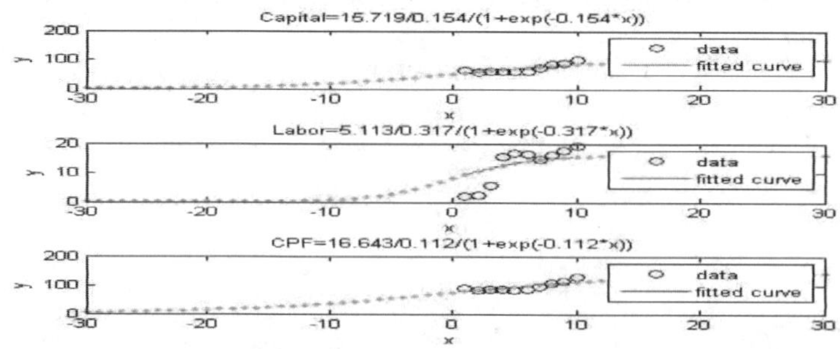

Figure 4. The logistic primary curve without the time period shift

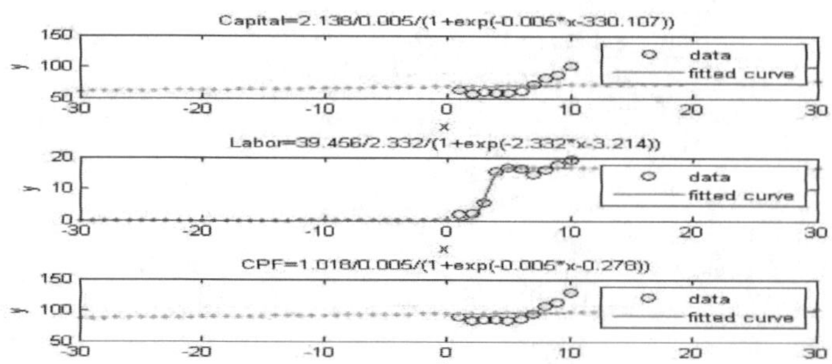

Figure 5. The logistic primary curve with the time period shift

The coefficients of the first phase are presented in table 3.

Table 3. Coefficients of the modelled production factors

	Primary curve without time period shift	Primary curve with time period shift
Capital	K = 15.719	K = 2.138
	a = 0.154	a = 0.005
Labour	K = 5.113	K = 39.456
	a = 0.317	a = 2.332
CPF	K = 16.643	K = 1.018
	a = 0.112	a = 0.005

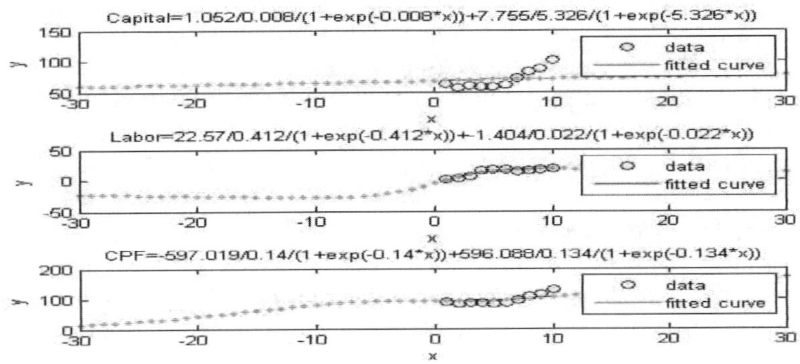

Figure 6. The logistic second-degree curve without the time period shift

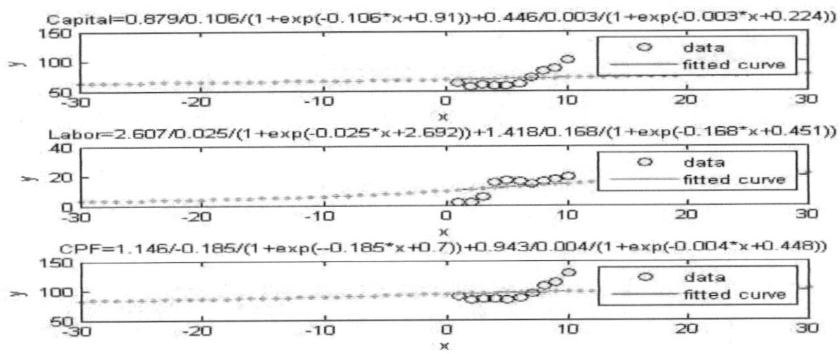

Figure 7. The logistic second-degree curve with the time period shift

The coefficients of the second phase is presented in table 4.

Table 4. Coefficients of the modelled production factors

	Second-degree curve without time period shift	Second-degree curve with time period shift
Capital	K = 1.052	K = 0.879
	a = 0.008	a = 0.106
Labour	K = 22.57	K = 2.607
	a = 0.412	a = 0.025
CPF	K = 597.019	K = 1.146
	a = 0.14	a = 0.185

The results of modeling have shown that the change from the multiplicative dependence to the logistic dependence have the confirmation especially with the labor factor. Concerning the cognitive production factors and the capital the logistic curve is not so evident. Perhaps the data series are not too long and we need to enlarge it. Another possible explanation of weak statistical meanings is the direct correlation between theoretical notions of production factors and real economic indicators. As this work is in progress we consider the given results rather meaningful and the future work will be devoted to find the indirect parameters of cognitive production factors' influence.

5. Conclusion

In the literature of the production processes the dominant idea is standard multiplicative production functions. However they don't take into account in the present time the manufacturing terms have changed dramatically: Industry 4.0, expansion of IT, etc. In the present paper an attempt has been undertaken to consider cyber and intelligent technologies by means of introduction a new notion – cognitive production factors. The review of the traditional approaches for the efficiency modelling has shown that they don't allow taking into consideration the technologic ceiling of existing production factors and the necessity to transfer to other technologic base. In the present paper the new approach has been elaborated and production function has been modified on the base of the logistic curve. The modelling of the modified production function on the open data has allowed evaluating the statistical meaning of the suggested idea. Though the modelling results are quite modest from the point of statistics, we consider them successful. The present paper presents a part of a larger scientific work and enlarged by other aspects and results. In the end it is necessary to mention about the limitation of this paper. Certainly the data series is not too large and probably we should find other quantitative equivalents of such qualitative notion as cognitive production factors.

References

[1] E. Gorlacheva, Cognitive production factors: the problem of basic notions, Az. Scien.Res.,2 (2018)
[2] E. Bukhvalov, The formation of the high-tech sector in Russian economy, ed. by A. Tatarkin. Ekateringburg, 2016.
[3] E. Gorlacheva, Cognitive production factors: a research problem statement, Harald of MSRU, 2 (2018)
[4] M. Mamonov, A. Pestova, E. Sabelnikova, A. Apokin Approaches to production factors evaluation and technology development of national economy; the review of world practice, Forec.Prob., 6 (2015)
[5] R. Nizhegorodzev, Logistic modelling of economic dynamics, Manag. Issues, 3(2004).
[6] R. Barro, X. Sala-i-Martin, Technological Diffusion, Convergence and Growth, Jour. Econ.Growth, 2 (1997)
[7] R. Capolupo, The New Growth Theories and Their Empirics after Twenty Years, The Op.Acc., Op.-Ass.E-Journ. 3(1), (2009)
[8] L. Castillo, D. Salem, J.Guasch, Innovative and Absorptive Capacity of International Knowledge: an Empirical Analysis of Productivity Sources in Latin American Countries. The World Bank Pol.Res.Work.Pap, 5931, (2012)
[9] E. Nazrullaeva, The evaluation of technology progress level in Russian economy, Quant., 5 (2008)
[10] N. Mankiw, D. Romer, D. Weil, A Contribution to the Empirics of Economic Growth, Quart. Journ. Econ., 107, (1992)
[11] W.J. Meeusen, J. Broek van der, Efficiency Estimation from Cobb-Douglas Production Functions with Composed Error, Int.Econ.Rev, 18 (1977)
[12] P. Duglas, Do the manufacturing laws exist? In Book of International economy, ed. by A. Kireev, M. (2006)
[13] E. Gorlacheva, M.Kutz, Modelling of the efficiency in the high-tech enterprises, Econ.Scien, 10 (2018)
[14] E. Gorlacheva The elaboration of a modified logistic curve, Innov. in Manag., 2 (2018)
[15] M. Bendikov, I. Frolov, High-tech sector of Russian Industry: state, tendencies, mechanisms of

the innovation development. M. (2007)
[16] A. Arutyunova, E. Smirnova, Y. Isakova, V. Zelik, M. Kozin Social and Economic Space of the Territory: Issue: Development Conditions, Structure, Evaluation Criteria, Inter. Jour. App. Bus. Econ. Res., 23 (2017).
[17] E. Gorlacheva, A. Gudkov, I. Omelchenko, P. Drogovoz, D. Koznov Knowledge management capability impact on enterprise performance in Russian high-tech sector, Proc. Intern. Conf. Engin., Techn. Inn., (2018)
[18] E. Gorlacheva, A. Gudkov, D. Koznov, I. Omelchenko, The knowledge management capability of high-technology enterprises, Proc. 9th Intern. Joint Conf. Knowl. Disc., Knowl. Engine. and Knowl. Manag. (2017)
[19] J. Bhagvati, Destroying growth, in Book of International economy, ed. by A. Kireev, M. (2006)
[20] P. Johns, The correlation between factors and theorem of Heckscher-Ohlin, in Book of International economy, ed.by A. Kireev, M. (2006)
[21] L. Grebnev, Resources and Factors: similarity, difference or contrast? Econ.Issues, 4 (2010)
[22] A. Chursin, P. Drogovoz, T. Sadovskaya, V. Shiboldenkov, The Dynamic Model of Elements' Interaction Within System of Science-intensive Production Under Unstable Macroeconomic Conditions, Jour. App. Econ. Scien., 5(51), (2017)
[23] A. Chursin, P. Drogovoz, T. Sadovskaya, V. Shiboldenkov, A linear model of economic and technological shocks in science-intensive industries, Jour. App. Econ.Scien., 6(52), (2017).
[24] V. Melikhovsky, Actual issues of production factors' theory. Yaroslavl (2013)
[25] V. Bilgram, A.Brem and K.I. Voigt, "User-centric innovations in new product development: systematic identification of lead users harnessing interactive and collaborative online tools," Intern.J. Innov. Manag., vol.12, 2008
[26] V. Shiboldenkov, M. Dobrinetz, Economic and mathematical modelling of innovation diffusion in manufacturing-distributive systems, Proc. Charn.Read., 2017.
[27] A. Dynkin, M. Grachev, A. Dagaev, N. Ivanova, Innovation development contours of world economy, M. (2000)
[28] A. Balashov, Y. Martyanova, Reindustrialization of russian economy and development of high-tech sector, Econ. Iss., 9 (2015)
[29] V. Vlasova, L. Gochberg, T. Kuznetsova, V. Rud, Innovation production output by industrial enterprises, in Sci., Tech., Inn., Inst.stat. res. Know. Econ., (2017)
[30] N. Komkov, M. Krotova, Innovation and technology factors perspectives of economy development on the base of resource-innovation strategy, Acad.papers of Inst.Nat. Econ. Forec. 12 (2014)
[31] I. Frolov, M. Bendikov, I. Ganichev, O. Koshovetz, Analysis of state financial possibilities for high-tech sector development, For. Rus. Techn. Moder., 1, (2010).
[32] Development strategy of electronic industries in Russia, http://www.elcomdesign.ru/netcat_files/File/Gosprogramma_REiRP-1.pdf
[33] Development strategy of shipbuilding in Russia, http://federalbook.ru/files/OPK/Soderjanie/OPK-10/V/Strategiya.pdf
[34] I. Frolov, Innovation as a process of capital extended development: is it possible to build innovation economy in Russia?, J.Econ. & Busin., 6, (2012)
[35] Open Data of a high-tech enterprise, www.kamaz.ru

IOP Conf. Series: Materials Science and Engineering **630** (2019) 012021 doi:10.1088/1757-899X/630/1/012021

Transient thermal contact in gyro unit–platform assembly

Mikhail V. Murashov

Bauman Moscow State Technical University, 105005 Moscow, Russia

murashov@bmstu.ru

Abstract. Thermal contact conditions of parts change in time in the process of heating. Cases of thermal contact instability may occur even under constant boundary conditions. This is due to the combination of temperature and strain fields interdependence based on thermal expansion, the influence of shape of parts and the conditions of their connection in the assembly. In this paper, using the example of a model problem of thermal contact of a gyro unit–platform assembly, we study qualitative and quantitative aspects of the temporal behavior of thermal contact conductance. A significant qualitative difference between thermal contact conductance and its steady-state value has been obtained. Given that the magnitude of the deviation is within the acceptable engineering accuracy, it is of interest in terms of thermal contact model development.

1. Introduction

Heating leads to spatial thermal expansion of parts. This expansion in combination with the constraints of connections in the assembled structure leads to increased stresses and additional deformation. Thus, both tightening and gapping of contacting parts are possible.

There are possible contact options, when the thermal contact conductance does not reach its steady-state value for a long time or even never reaches one if it has an oscillatory character. This is possible if thermal expansion makes the contact open. For example, the controlled process of mechanical contact opening can be seen in widespread constructions of thermostats and thermoregulators. Figure 1 shows possible simplified construction of systems, where the transient process may fluctuate.

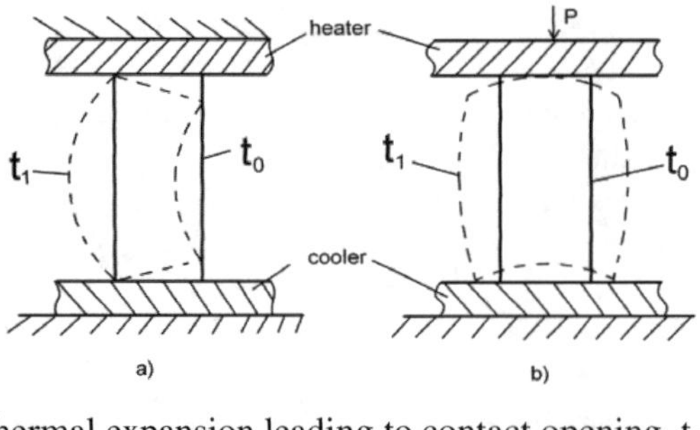

Figure 1. An example of thermal expansion leading to contact opening. t_0 is the shape of the body at the initial temperature, t_1 is the deformed warmed body with the open contact: a) the contact opening upon buckling due to thermal expansion; b) opening due to the deformation of the samples in the thermal contact conductance measurement test.

In real structures of complex shape there will be an intricate and often unpredictable surface warp, also influenced by dependence on heat fluxes and compliance resulting from the asperity deformation. Such a warp should be taken into account by solving transient problems with finite element models.

The temporal change in contact conditions leads to the change in thermal contact conductance. This changes the temperature field, which in its turn has an impact on thermal contact conductance as well. The latter, apparently, can explain the significant change in the measured dependence of thermal contact conductance on the strain rate during hot forging of metal described in [1]. Faster loading forms different temperature fields of the billet and the die and, therefore, changes the contact thermal conductance.

There are few works on transient modeling of thermal contacts, for example, [1-4]. In [5], for a one-dimensional model, it was shown that in the case of heating with short impulses, even a small thermal resistance has a significant effect on the temperature field.

The present study aims at determining the nature of thermal contact conductance and estimating its temporal deviation from the steady-state value. This is based on a model problem of gyro unit–platform assembly.

We will consider a calculation of the temperature field for gyro unit–platform assembly as a typical member of a platform inertial navigation system which significantly influences its temperature state [6].

In fact, thermal contact conductance varies at different points of the contact and to solve the problem we calculate the dependence of thermal contact conductance on contact pressure with our column micromodel of rough surfaces contact [7]. Then, this dependence is transferred to the finite element model of the gyro unit–platform assembly.

2. Description of gyro unit–platform model and formulation

In the right-handed Cartesian reference system, let us consider the hypothetical construction of a gyro unit attached to a platform (see Figure 2).

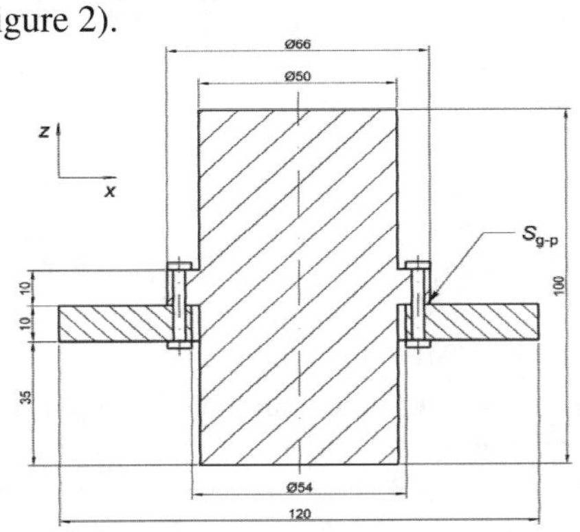

Figure 2. Geometry of macromodel (cross section shown, dimensions in mm). Sg-p is the contact surface of gyro unit–platform.

The model of the gyro unit is a solid cylinder with a flange, made of AISI 1020 steel. Inside the gyro unit we assume the presence of a thermostatic system maintaining on a constant level the temperature of some internal part of the device. At the same time, the temperature of the device body is not constant and is determined, among other things, by interaction with the platform. The model does not consider the internal parts and the thermostatic system as separate bodies, but it takes into account that this system generates heat with a power of 20 W, corresponding to a specific heat generation rate q = 94832 W/m^3 for the given volume of the gyro unit. Heat transfer by radiation is neglected. The gyro unit is fixed from the flange onto the square platform of 1050 UNS A91050 Aluminum with four M3 bolts made of AISI 5140 steel. Half-length of the platform side L_1 = 0.06 m. Distance from the bottom surface of the gyro unit to the bottom surface of the platform L_2 = 0.035 m. The behavior of the bolt material is considered linearly elastic. The clamp force of each bolt F is 2000 N. The joints of the parts are in a vacuum. Isotropic dry friction according to the Amontons-Coulomb law with a friction coefficient μ = 0.5, constant in the process of deformation, is taken into account in the joints. Initially, the contact surfaces of the macromodel are ideally flat. The nominal contact area of the gyro unit–platform is $1.103 \cdot 10^{-3}$ m^2. The model material properties are presented in Table 1.

The gyro unit–platform assembly model is constrained at the four corners of the lower platform plane as follows. One corner is constrained from displacement along all three axes x, y, z. The corner adjacent to it is constrained along axis z and along the axis orthogonal to the connection line of the first and second corners. The third and fourth corners are constrained only along axis z. Thus, these constraints do not oppose thermal expansion, i.e., do not cause additional thermal stresses.

Table 1. Material properties of the model.

Part	Gyro unit	Bolts	Platform
Material grade	AISI 1020 steel	AISI 5140 steel	1050 UNS A91050 aluminum
Modulus of elasticity *E*, GPa	212	212	71
Poisson ratio ν	0.36	0.3	0.32
Yield strength, MPa	280	-	55.7
Linear coefficient of thermal expansion α, 1/°C	$11.1 \cdot 10^{-6}$	$12 \cdot 10^{-6}$	$23.5 \cdot 10^{-6}$
Thermal conductivity *k*, W/(m·°C)	86	41	210
Specific heat capacity c_p, J/(kg·°C)	461	466	858
Mass density ρ, kg/m^3	7850	7810	2710

In this model, we assume the absence of axial and bending external forces that may affect the area of actual contact [8].

On the right wall of the platform, heat is removed with an intensity of h = 500 W/(m^2·°C) and coolant temperature $T_{coolant} = 20\ °C$. Thermal expansion was taken into account when calculating the assembly model. Reference temperature is $T_0 = 20\ °C$. The coupled problem of heat conductance and deformation in a static setting was solved. All loads were applied simultaneously in a single load step.

Let us move on to the index designations of axes of the Cartesian reference system $x_i, i = 1,2,3$. The following mathematical model corresponds to the problem, including equilibrium equations (1), the generalized Hooke's law (2), the flow rule (3), thermal strain relation (4), strain-displacement relations (5), von Mises yield condition (6), ratio for calculation of contact pressure of augmented Lagrangian method (7), Amontons-Coulomb friction law (8) on the contact surface, structural boundary conditions (9)-(12), heat equation (13), initial condition (14), and thermal boundary condition (15)

$$\sigma_{ji,j} = 0; \tag{1}$$

$$\varepsilon_{ij} = \frac{1+\nu}{E}\sigma_{ij} - \frac{\nu}{E}\delta_{ij}\sigma_{kk}; \tag{2}$$

$$d\varepsilon_{ij}^P = s_{ij}d\lambda; \tag{3}$$

$$d\varepsilon_{ij}^T = \alpha(T - T_0); \tag{4}$$

$$\varepsilon_{ij} = \frac{1}{2}(u_{i,j} + u_{j,i}); \tag{5}$$

$$(\sigma_1 - \sigma_2)^2 + (\sigma_2 - \sigma_3)^2 + (\sigma_3 - \sigma_1)^2 = 2\Phi(\varepsilon^P)^2; \tag{6}$$

$$p(x_i) = K\delta + \lambda_c,\ x_i \in S_c; \tag{7}$$

$$\tau_c \leq \mu p(x_i),\ x_i \in S_c; \tag{8}$$

$$\sigma_{33}^n(x_3) = \frac{F}{A_n},\ x_3 \in S_n^b, n = 1 \ldots 4; \tag{9}$$

$$u_i|_{x_i=(L_1,-L_1,L_2)} = 0; \tag{10}$$

$$u_1|_{x_i=(-L_1,-L_1,L_2)} = 0, u_2|_{x_i=(-L_1,-L_1,L_2)} = 0; \tag{11}$$

$$u_3|_{x_i=(-L_1,L_1,L_2)} = 0, u_3|_{x_i=(L_1,L_1,L_2)} = 0; \tag{12}$$

$$c_p\rho\frac{\partial T}{\partial t} - k\Delta T - q = 0; \tag{13}$$

$$T = T_0, t = 0; \tag{14}$$

$$-k\frac{\partial T}{\partial x_1} = h(T - T_{coolant}),\ x_i \in S_{conv} \tag{15}$$

where σ_{ij} and ε_{ij} are the Cartesian components of tensors of stress and strain, u_i are the components of displacements vector, δ_{ij} is the Kronecker delta, s_{ij} are the components of stress deviator tensor, *T* is the temperature, $\sigma_1, \sigma_2, \sigma_3$ are the principal stresses, λ, λ_c are Lagrange multipliers, $\Phi(\varepsilon^P)$ is the function of the material's stress-strain curve, *p* is the contact pressure, *K* is the contact stiffness, δ is the contact gap size, τ_c is the tangential stress on the contact surface, S_c is the surface of contacts, *n* is a bolt index, S_n^b is a cross section of bolt *n*, A_n is an area of S_n^b, Δ is the Laplace operator, S_{conv} is the cooled platform surface.

The problem is solved using ANSYS finite element software. For the simulation of bolts, pretension type of analysis and PRETS179 finite elements are used. The finite element mesh of the assembly model is shown in Figure 3.

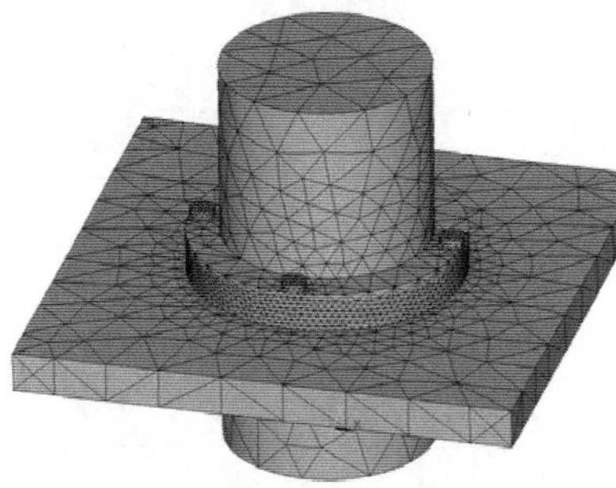

Figure 3. Finite element mesh of the assembly model.

To determine the contact pressure and gaps for all contact areas, the Standard contact type is selected in the ANSYS options. Augmented Lagrange method is used as a contact algorithm with Gauss point contact detection method.

On the contact finite elements of the macromodel thermal contact conductance is specified as the TCC parameter of ANSYS. In the assembly, it is possible to define the contact of the gyro unit – platform, which is the most influential in the heat transfer, and the contacts of the gyro unit – bolts and the platform – bolts. For two latter contacts the TCC parameter is estimated to be 100,000 W/(m$^2\cdot$°C).

Since the bolts form a field of contact pressures markedly different from uniform, the thermal contact conductance is specified tabularly as a function of pressure TCC(*p*) where *p* is the contact pressure on a specific finite element. This dependence is obtained from the calculation of the column micromodel of contact rough bodies with dimensions 90×90×1125 μm described in [7], taking into account the indentation size effect (ISE).

The calculations were carried out within the approach of small displacements, since, according to the results of preliminary calculations, the inclusion of large displacements led to an insignificant change in the result.

The temperature field of the macromodel was determined, and then, the averaged over the nominal area thermal contact conductance of the gyro unit – platform contact region was calculated by the relation

$$\overline{\alpha_c} = \frac{\bar{q}}{\overline{T_2} - \overline{T_1}}$$

where $\bar{q}$ is the average heat flux density over the area of finite elements of the lower nominal surface, $\overline{T_1}$, $\overline{T_2}$ are the average temperatures over the area of finite elements of the nominal lower and upper contact surfaces, respectively.

The model takes into account the effect of thermal expansion on the real contact area. Because of the change in shape of the cylindrical body of the gyro unit, tangency takes place in the form of a narrow ring along the outer edge of the nominal contact area. Also, areas near the bolts are in direct contact.

The selected interval of the calculations ranges from 0 to 7000 s which is of interest from the engineering point of view for similar constructions of platform inertial navigation systems.

3. Numerical simulation and results

The calculation results have shown that the maximum stresses in the structure do not exceed the yield strength. The thermoelastic effect, where the redistribution of stresses can lead to a temperature change, is neglected, since even at the pressure of 1-2 GPa it can result in a short-term temperature alteration only by 0.1-1% [9]. Therefore, the coupled thermomechanical problem has been solved without taking thermoelastic damping into account.

When solving a transient problem, the Newmark model and the *High Speed* time integration mode are used. Stepped loading is used and in order to avoid solution oscillations, a sufficiently small time step was taken in accordance with Table 2.

Table 2. Selection of time step for a transient calculation.

Time period, s	Time step, s
0.001-0.1	0.001
0.1-0.2	0.002
0.2-20	0.01
20-320	1
320-700	2
700-3000	10
3000-50000	100

With small time steps, it is especially difficult to calculate the first step. The solution requires a large number of intermediate iterations (several dozens).

The results have been compared with the steady-state solution, which we obtained earlier in [6] for the same formulation. The calculated thermal contact conductance of the gyro unit–platform connection averaged by the nominal area was α_c = 20739 W / (m^2 · °C). The maximum temperature of the gyro unit was 61.53 °C.

Before the heating starts, the bolts are tightened and the parts are pressed together. At this moment, the thermal contact conductance equals infinity, since the structure has uniform temperature field and the temperature jump of the contact equals 0. During the first time step, the instantaneous value of thermal contact conductance, which is usually not far from the steady-state one, is determined. It grows over time, reaches its maximum and decreases below the steady-state value (see Figures 4, 5). Then it grows reaching the second maximum and decreases to the value which exceeds the steady-state one by about 5% (see Figure 6). The first maximum, which becomes apparent at a small time step, is explained by the proximity to the singularity that results in temperatures less than T_0 in the numerical solution. The appearance of the second maximum during heating is characteristic of flat–flat contacts

[10] and is related to thermal expansion of parts. The final value obtained in the solution differs from the steady-state one due to the errors of the finite element method.

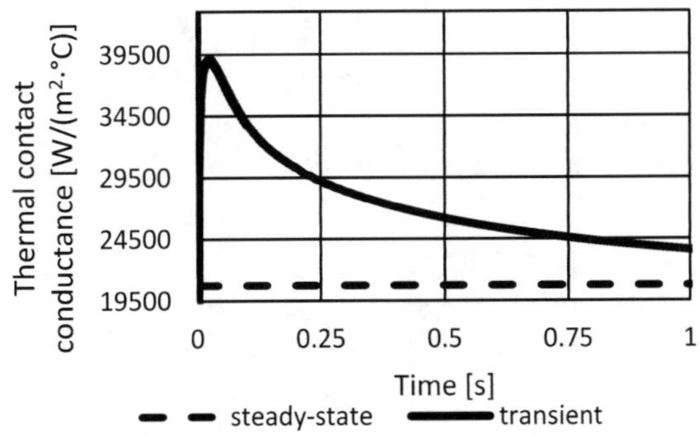

Figure 4. The change of the thermal contact conductance of the gyro unit–platform assembly(up to 1 s)

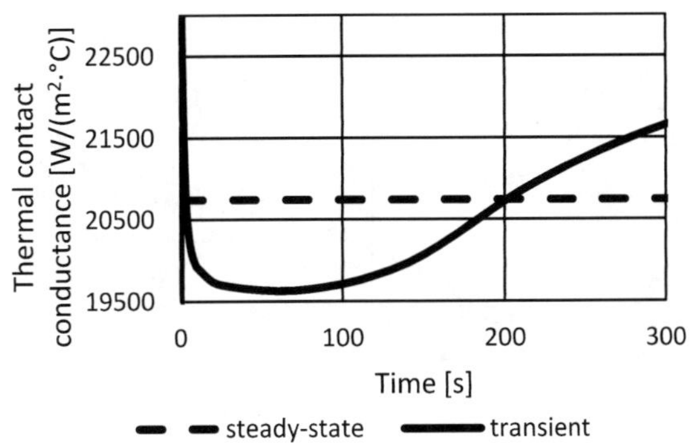

Figure 5. The change of thermal contact conductance of gyro unit–platform assembly (up to 300 s).

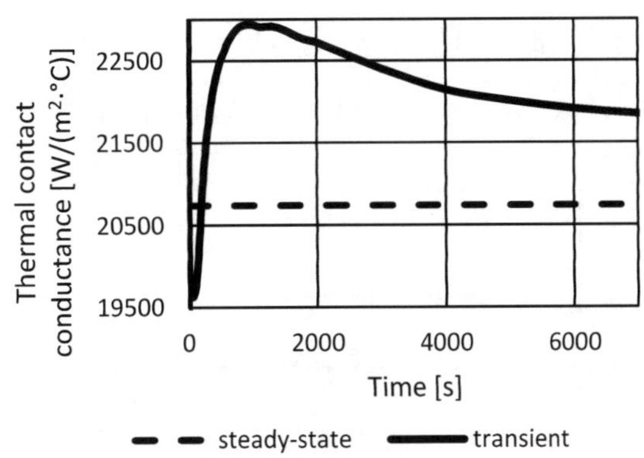

Figure 6. The change of thermal contact conductance of gyro unit–platform assembly (up to 7000 s).

The fluctuations in thermal contact conductance on the interval of 20–7000 s give a deviation from the steady-state value from –6% to +11%. This deviation leads to a change in the temperature field of this assembly model by less than 0.1 °C, which can be neglected in engineering practice. Nevertheless, it may be noticeably larger in other contacts and, moreover, this deviation is significant from theoretical

IOP Conf. Series: Materials Science and Engineering **630** (2019) 012021 doi:10.1088/1757-899X/630/1/012021

point of view as it shows a considerable contribution of transience to the thermal contact model. It should be noted that the deviation could be substantially greater with smaller bolt clamping force.
Figures 7-9 show the variation of heat flux, averaged temperatures in the contact and the temperature jump versus time. The form of these variations depends entirely on the size and configuration of the parts in a contact, the thermophysical and mechanical properties of materials, and the conditions of the contact. Apparently, the variations change monotonously and even in such conditions there is a significant change in thermal contact conductance.
The thermal contact conductance of this contact practically reaches its steady-state value within the taken period of 7000 seconds.

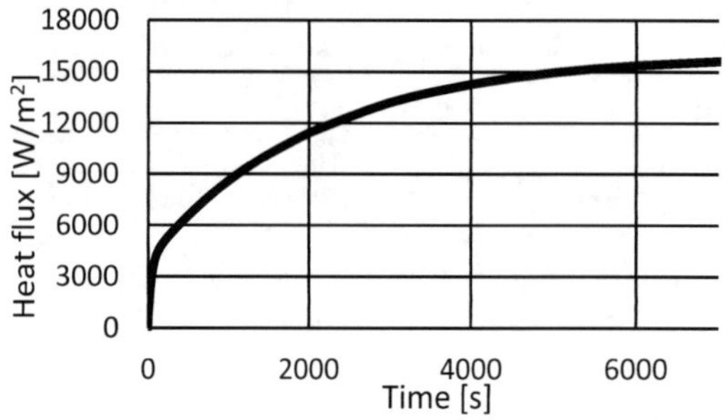

Figure 7. The change of heat flux versus time.

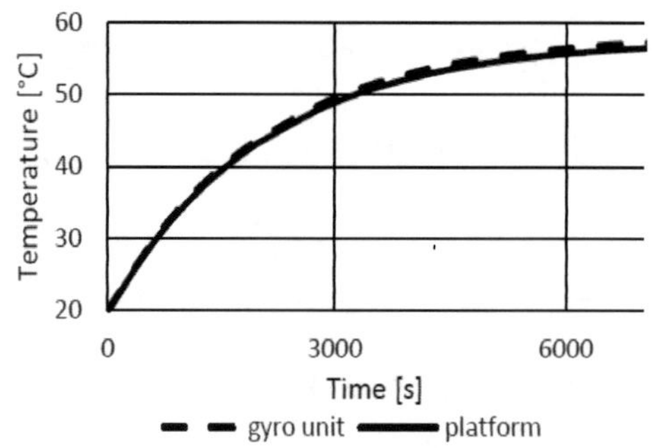

Figure 8. Average temperatures of the surfaces in contact versus time

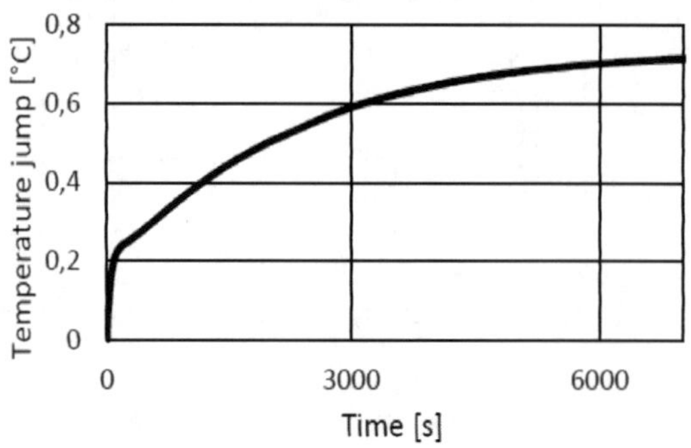

Figure 9. Temperature jump versus time.

4. Conclusion

It has been determined that the thermal contact conductance in the process of even heating of a structure varies considerably, both exceeding its steady-state value and going below it. In engineering practice, the transient behavior of responsible contacts can be analyzed by means of commercial finite

element software provided that we have the dependence of thermal contact conductance on contact pressure. The widespread approach of calculation using only one constant contact-specific value of thermal contact conductance is insufficient in this case.
The obtained deviation of 11% is of great value for the developers of thermal contact models as it places consideration of transience among significant processes in modeling.

References

[1] B.S. Lu, L.G. Wang, Y. Huang, Effect of deformation rate on interfacial heat transfer coefficient in the superalloy GH4169 hot forging process. *Appl. Therm. Eng.* 2016, **108**: 516-524.

[2] J.V. Beck, Determination of Optimum Transient Experiment for Thermal Contact Conductance. *Int. J. Heat Mass Trans.* 1969, **12** (1): 621-633.

[3] T. Loulou, E.A. Artyukhin, J.P. Bardon, Estimation of Thermal Contact Resistance during the First Stages of Metal Solidification Process: II-Experimental Setup and Results. *Int. J. Heat Mass Trans.* 1999, **42:** 2129-2142.

[4] C. Fieberg, R. Kneer, M. Korthauer, E. El-Magd. Transient contact heat transfer coefficients from infrared temperature measurements. *Proc. of the International Heat and Mass Transfer Conference IHTC13.* 2006:12.

[5] A.D. Barber, J.H. Weiner, B.A. Boley, An analysis of the effect of thermal contact resistance in a Sheet-stringer structure. *J. Aeronaut. Sci.* 1957, **24** (3): 232-234.

[6] M.V. Murashov. Temperature Field Simulation of Gyro Unit-Platform Assembly Accounting for Thermal Expansion and Roughness of Contact Surfaces. In: Hu Z., Petoukhov S., He M. (eds.). *Advances in Artificial Systems for Medicine and Education II.* 2020, **902**: 601-616.

[7] M.V. Murashov, S.D. Panin, Numerical modelling of contact heat transfer problem with work hardened rough surfaces. *Int. J. Heat Mass Transf.* 2015, **90**: 72-80.

[8] I. Coria, I. Martín, A.-H. Bouzid, I. Heras, M. Abasolo. Efficient assembly of bolted joints under external loads using numerical FEM. *Int. J. Mech. Sci.* 2018, **142-143:** 575-582.

[9] V.L. Hilarov, A.I. Slutsker. Description of the thermoelastic effect in solids in a wide temperature range. *Phys. Solid State.* 2014, **56**: 2493-2495.

[10] S. Kumar, A. Tariq. Determination of thermal contact conductance of flat and curvilinear contacts by transient approach. *Exp. Therm. Fluid Sci.* 2017, **88**: 261-276.

IOP Conf. Series: Materials Science and Engineering **630** (2019) 012022 doi:10.1088/1757-899X/630/1/012022

Sensor subsystem design for small unmanned surface vehicle

Georgii Kolev, Ekaterina Solomevich, Ekaterina Rodionova, Ekaterina Kopets and Vyacheslav Rybin

St. Petersburg Electrotechnical University "LETI", Saint Petersburg, Russia

gyukolev@stud.etu.ru

Abstract. The construction of unmanned vehicles is a major trend in a modern robotics and automation. One of the promising branches is the development of unmanned boats and surface vehicles (USV). They possess the great variety of possible applications, including environmental studies, military and security issues. In this paper, we consider the design and rapid prototyping of sensory subsystem for small-sized USV. The proposed sensor system incorporates the lidar, Doppler radar rangefinder and video camera in a waterproof housing. The housing is made with additive technology and provides the efficient and rigid placement of chosen set of sensors. The circuitry and design of the sensory subsystem are given and discussed in comparison with existing solution. The prototype of the system is described and the results of its experimental testing are provided.

Introduction

Nowadays robotics is one of the most dynamic and relevant branches of technical knowledge. It is believed that the development of unmanned surface vehicles (USV) s will enter a new phase in the near future, as they could soon be applied widely in both military and civilian service [1]. During the design of the floating robot, the developer decides how to organize the interaction between the robotic system and the environment; he also decides how to obtain information about robot location and the parameters of the surrounding space. In order to adjust the trajectory of motion, avoid the obstacles and accurate positioning of the manipulators, the robot must have a kind of "sense organs" [2]. Thus, the sensory subsystem is a block of a robot, which converts the physical parameters of the environment into information. For the efficiency and reliability of the sensory system of an unmanned boat, it is necessary to fulfil such requirements as moisture protection, clear visibility in rough and choppy water, low weight. The subsystem and its elements must be placed in the way to lower the center of gravity for maximum stability of the boat and have streamlined structural elements to reduce air resistance and minimize the risk of structural failure in case of large waves. All these issues became extremely important in a case of small USV, where they are accompanied by the low energy consumption and weight economy. Nevertheless, the overall requirements for sensor systems of small USVs are not significantly lower, than for full-size unmanned ships. We still need to solve positioning and obstacle avoiding problems, but are limited with resources. This makes the development of new design techniques relevant for small USV construction.

Sensory system design

In our design, we chose the following set of sensors: a Doppler radar, a lidar and a camera. For the subsystem of early detection, the lidar and the radar are chosen. The lidar performs accurate measurements at a sufficient distance, but some properties of the obstacles may confuse its readings. In order to avoid this, the radar (Fig. 3) is placed coaxially, allowing to detect the objects with complex reflective properties. These sensors are installed on a radially movable platform driven by a stepper motor.

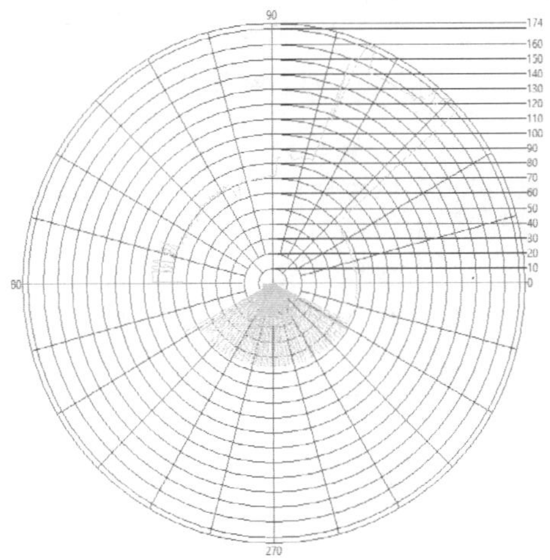

Figure 1. Map of distances to the nearest objects

For accurate sensor readings and a stable camera image, the stabilization device is necessary. It is supposed to use an active stabilizer on which the lidar and camera will be placed. For high performance, they both require to be relatively precise positioned. Since these sensors have a high acquisition speed, they allow us to construct a precise circular (360 degrees) map of distances to the nearest objects (Fig. 1). The minimal step of the selected stepper motor is 0.9 degrees. The lidar has the longest response time in the rotary sensory head of 0.02 seconds. Therefore, it is necessary to perform 400 motor steps to complete a full 360-degree rotation. On each step the sensory system measures the distance during 0.02 sec. Taking this into account, the system will require 8 seconds for a full scan. This time can be reduced by dropping the number of motor steps and, subsequently, the scanning resolution. Despite the variety of possible sensors available on market, in our project we chose non-expensive and available sensors that meet minimal accuracy criteria and can be easily purchased for mass-production of USV.
The first sensor to choose was the Lidar-Lite v3 (Fig. 2) because of its sufficient measurement accuracy and the required range.

Figure 2. Lidar-Lite v3 [3]. Characteristics: unit size (HxWxD): 40x48x20 mm (1.6"x1.9"x0.8"); weight: 22 g (0.77 oz); accuracy: +/–2.5 cm at distances greater than 1 meter. range: 5 cm to 40 meters; update rate: up to 500 Hz (0.02 sec).

The radar is the HB100, a compact, powerful sensor with low power consumption (Fig. 3).

IOP Conf. Series: Materials Science and Engineering **630** (2019) 012022 doi:10.1088/1757-899X/630/1/012022

Figure 3. Microwave Motion Sensor Module [4]. Characteristics: frequency setting: 10.520–10.530 GHz; settling time: 3–6 μSec (0.000006) sec; received signal strength: 200 μVp-p; noise: 5 μVrms; antenna beam-width (3 dB) – azimuth: 80°; antenna beam-width (3 dB) – elevation: 40°; supply voltage: 4.75–5.25 VDC.

The PI Cam. V2.1 was chosen as a main camera because it has small dimensions, low cost and relatively high performance (Fig. 4).

Figure 4. PI Camera V2.1 [5]. Characteristics: maximum resolution: 8 MP (3280×2464); supported video formats: 1080p (30fps), 720p (60fps), 640×480p (90fps); lens aperture: f/2; size: 2592×1944

To move the sensor system we chose the bipolar two-phase stepper motor 36HT20-0504MA (Fig. 5). This motor allows making turns at an arbitrary angles that are the multiples of 0.9°. The motor has four inputs: two for each winding of the available two phases.

Figure 5. Stepper motor [6]. Characteristics: step: 0.9 ° ± 5% (400 per turn); rated supply voltage: 6.5 V; holding torque: not less than 0.95 kg × cm; maximum starting speed: 1500 steps / sec; shaft diameter: 5 mm; weight: 0.16 kg.

To control the motor using a microcontroller, one needs an intermediary stepping motor driver. We choose Troyka-Stepper Motor Driver for our project since it is ready-to-go and easy-to-use solution based on the L293D driver chip (Fig. 6).

Figure 6. Motor Driver [7]. Characteristics: motor power supply: 4.5–25 V; peak voltage on Vin contacts: 35 V; voltage of the logical part: 3.3–5 V; continuous current: up to 600 mA; peak current: 1200 mA.

Figure 7. Sensor subsystem ranges of detection.

Figure 7 shows the sectors and ranges for the selected sensors. Due to the fact that the sensors are fixed on the rotary mechanism, the whole area within a radius of 35-40 meters can be covered. Of the three sensors available, the lidar has the narrowest viewing angle, the camera is the largest.

Housing design

The microwave radar module, camera and lidar are placed on the rotary head (Fig. 8) that can only turn 170 degrees in each direction. The real angle is limited because of the restrictions determined by wire flexibility. Such construction allows to scan the wide operational area and to create a distance map.

In our design we implemented model-based principles, creating a set of digital twins before real construction. One can see from Fig.8, that the camera is located on top, while the radar is located beyond lidar position. The casing does not affect measurements since it does not create a difference between the output and the input radar signals. Since microwaves overcome an obstacle in the form of a casing, it is not necessary to design special holes for it.

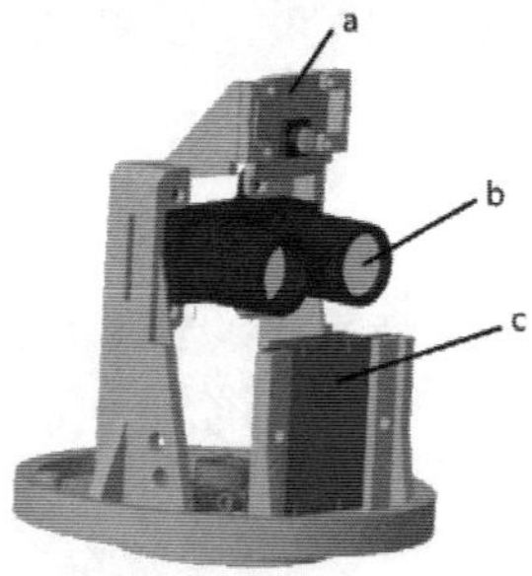

Figure 8. Rotary head: a – PI camera, b – lidar, c – microwave sensor module.

The overall height of the construction is 112 mm, the width is 100 mm. This construction is covered with the protective casing (Fig. 9), which provides the necessary dust and moisture protection. The casing is secured with four M3 screws. The head is fastened to the shaft by clamping details. The clamp is attached to the head with two M5 screws.

Figure 9. The cutaway of sensory head in protective casing.

Thus, the total height of the rotary head with the protective casing is 130 mm and width is 105 mm. The inner part of the base is equipped with two bearings with an internal diameter of 8 mm. They make the shaft more resistant to various vibrations and increase the overall rigidity of the construction. The agility of the sensor head is provided by the radial ball bearing, which serves as an interlayer between the support base and the sensor head (Fig. 10), the outer diameter of the bearing is 63.5 mm, the internal diameter is 50.8 mm.

Figure 10. Base.

To prevent the ingress of water and dust into the structure through the head and base joint assembly, it is protected by a stuffing box with an inner diameter of 85 mm, an outer diameter of 100 mm, and a width of 10 mm.

The place of contact of the stuffing box with the movable part has a gasket in the form of a copper ring, which ensures the best fit of the stuffing box to the overall construction and increases its efficiency and service life. The smooth copper ring with the application of a lubricant oil will wear out the stuffing box much less in a case when the base body will be made of PLA plastic.

The shaft is attached to the motor through a flexible coupling that is able to damp some vibrations. Fig. 11 illustrates the whole assembly.

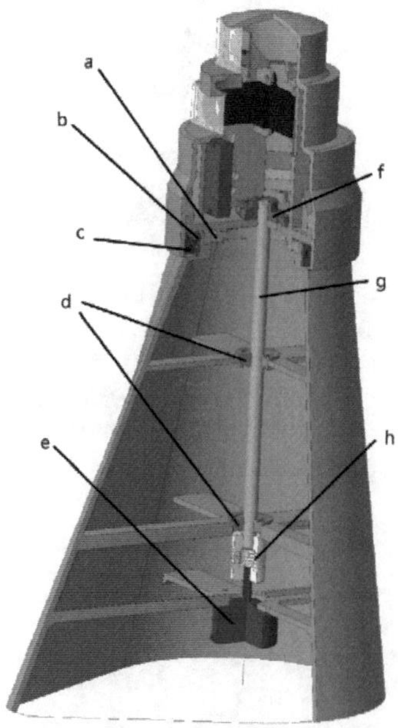

Figure 11. Overall construction (cutaway): a – radial bearing, b – copper ring, c – ring gasket, d – radial bearing, e – stepper motor, f – clamping of motor shaft, g – motor shaft, h – clutch.

The cables from the sensors and the motor are connected directly to the Raspberry Pi placed into the sealed container inside the boat's hull.

Figure 12. The prototype of USV with the sensor subsystem.

To demonstrate the whole structure, a model of the boat was developed with the sensor subsystem installed (Fig.12).

Conclusion

In this paper, the requirements imposed on the sensory system of the small unmanned surface vehicle were formulated. These requirements include accuracy, protection against atmospheric influences, scanning surface and stabilization. The concept of sensoric subsystem was developed and 3D models were done in order to estimate the overall applicability of given approach. The product is now on prototyping stage, and we plan to publish a larger follow-up paper about our USV in peer-reviewed journal after the development of the prototype.

The described sensors and other elements of this design are selected in accordance with the permissible weight, dimensions and power consumption, and the general configuration of sensor system has been synthesized. Other types of sensors such as thermal camera, stereoscopic vision systems are energy consumptive types of sensors. In our design we were restricted by funding limitations as well. We plan to consider other types of sensors in a further development of our project.

The use of different sensors in this system makes it possible to avoid the problems inherent in using a single sensor of any type. Installing sensors on a rotating stabilized platform allows to scan a broad sector of the environment and acquire more information about USV operational area. To rotate the platform with installed sensors the stepper motor was chosen.

As an improvement of this system, we plan to implement the signal processing system. Real data obtained from the sensors must be processed to clear the resulting signal from interference. We can also add permanently installed ultrasonic sensors to the system (for example, 4 sensors located at a 90-degree angle to each other) to increase system reliability in close-range contacts. It is also possible to place promising types of sensors, such as chaotic sonar, resistant to crosstalk and environmental noise [8].

References

[1] R. J. Yan, S. Pang, H. B. Sun, Y. J. Pang, JMSA, 9(4), 451-457. (2010)

[2] Sensor-based robot control [Electronic resource] / eecs.uottawa.ca (2013) – Access mode: http://www.site.uottawa.ca/~petriu/CEG4392-IntroRobotics-Sensors.pdf.

[3] E. Ayala, J. Sotamba, B. Carpio, O. Escandón, Proceeding of 2018 IEEE Third Ecuador Technical Chapters Meeting, 1–5 (2018)

[4] V. C. Nguyen, D. K. Dinh, Proceeding of 2014 International Conference of Advanced Technologies for Communications, 371–376 (2014)

[5] G. Senthilkumar, K. Gopalakrishnan, V. S. Kumar, IJETTCS, **3(2)**, 213–215 (2014)

[6] Warren G. Mastering the raspberry PI (2014)

[7] L293D Quadruple half-H drivers. Datasheet Catalog [Electronic resource] (2009) – Access mode: http://www.datasheetcatalog

[8] T. I. Karimov, D. N. Butusov, V. V. Gulvanskiy, D. V. Bogaevskiy, *Progress in Electromagnetics Research Symposium – Fall*, 2521–2527 (2017)

IOP Conf. Series: Materials Science and Engineering **630** (2019) 012023 doi:10.1088/1757-899X/630/1/012023

Roller Bearing Fault Diagnosis Based on Empirical Mode Decomposition and Targeting Feature Selection

Xiaoyue chen[1,2*], Dang Ge[1,] Xiong Liu[1], and Mengchao Liu[1]

[1] School of Electrical Engineering and Automation, East China Jiaotong University, Nanchang 330013, China
[2] Huazhong University of Science and Technology, Wuhan, Hubei 430074, China

*976932632@qq.com;_wdmxdefeng@126.com; 997921925@qq.com;_ 601075460@qq.com.

Abstract. At present, the feature extraction of frequency signal based on empirical mode decomposition (EMD) has been widely studied and applied in fault diagnosis of rolling bearings. However, there are still some shortcomings in fault diagnosis based on EMD. Therefore, a fault diagnosis method based on the combination of EMD and target feature selection (TFS) is proposed in this paper. The method firstly analyzes the fault signal through EMD and extracts the fault features. Then, it removes the redundant features and optimizes the feature subsets by using TFS. TFS selects the most effective feature for each target sample space through filtering evaluation criteria and heuristic search strategy, thereby effectively improving the accuracy and efficiency of fault diagnosis.

1. Introduction

Rolling bearing is a key component in rotating machinery such as industrial machinery, spaceflight and navigation. It needs high reliability and safe operation. However, due to heavy workload in a long and bad environment, mechanical fault often

occur, affecting industrial production, increasing cost and maintenance time. Therefore, the safety and reliability of rolling bearings are particularly important for mechanical equipment [1]. The timely diagnosis of rolling bearing fault is the key to ensure safety and reliability, so fault diagnosis has great research value and significance [2].

In the past few years, a variety of signal processing methods have been proposed to detect and diagnose bearing defects. These methods include vibration and acoustic measurements, temperature measurement and wear particle analysis [1,5], in which bearing vibration signals are still the most direct, simplest and most effective information for the analysis of bearing defects [2,5]. Currently, non-linear, non-Guess, and non-stationary processing methods in vibration signals include short-time Fourier transform, Local feature-scale decomposition (LCD), Wavelet transform(WT), etc. These methods are widely used in feature extraction of fault signals of rotating machinery [3-5]. Compared with these methods, the EMD proposed by Hilbert-Huang can accurately give the combination of the frequency and time distribution of the signal energy. The local features of the signal time and frequency domain are accurately expressed by the instantaneous frequency, so this method is more effective in extracting the fault feature signal [3,6]. Based on EMD, some signal processing methods such as Ensemble Empirical Mode Decomposition (EEMD) [7] and Generalized Empirical Mode Decomposition (GEMD) [8] are proposed.

After extracting bearing signal features, there are still invalid f redundancy feature which affects the accuracy of feature subset characterization. In order to solve this problem, effective feature optimization and selection are usually used. At present, the feature selection method can be divided into three categories according to the feature evaluation criteria: The Filter, the Wrapper, and the Embedded [9]. Based on these, using Dynamic Relevance and Joint Mutual Information Maximization (DRJMIM), the redundant features are combined with dynamic weights to reduce the possibility of selection of real redundant features [10]; Based on Affinity Propagation (AP) clustering, RReliefF and Sequential Forward Search, the ARSFS is proposed to select the effective subset of the original feature set, reducing the feature space dimension and multiple correlation [11]. In addition, Ben Ali proposes a feature selection method combining Linear Discriminant Analysis (LDA) with Principal Component Analysis (PCA) [5], and successfully applied to fault diagnosis. These feature selection methods have achieved good results in eliminating invalid features, but there are still some shortcomings: 1). The combination of clustering and search is generally adopted, which results in the complexity of feature optimization. 2). For a variety of fault categories, effective feature selection is more difficult. Therefore, in order to solve this problem, this paper proposes a target feature selection method based on the empirical mode decomposition. The method is based on the transformation of multi classifications into two classifications. After the fault defect is analyzed by EMD, finding the optimal feature between two kinds of fault categories, and the nonoptimal feature is eliminated. Through experiments, it can be seen that this method not only has a good diagnostic effect on various fault categories, but also has a simple feature optimization process.

In the process of fault diagnosis, an accurate classifier is very important. The accuracy of classifier recognition is determined by the quantity and quality of the input features. Therefore, the classifier should be selected according to the characteristics of the input features. Although the KNN classifier is simple in calculation, it has poor ability to reject invalid features [12-15]. Although the SVM classifier has an inhibitory effect on the feature of invalid redundancy, the structure is too complex and time-consuming, and it is not suitable for a large number of feature recognition [12-15]. Considering comprehensively, a Probabilistic Neural Network (PNN) with a simple structure and stable classification effect is used as a fault classifier [5-16].

The main work of this paper is as follows: 1) extract the vibration signal features of rolling bearing based on EMD. 2) Selecting effective feature sets from all feature sets based on TFS. 3) Using PNN to complete the diagnosis of rolling bearing.

2. Target Feature Selection (TFS)

Through the in-depth analysis of 9 kinds of faults of the rolling bearing, as long as we can accurately distinguish any two kinds of faults, we can accurately achieve accurate diagnosis of multiple faults. As shown in figure 1, identifying four different color samples of red, yellow, black and green, as long as any two different color samples are distinguished from each other, that is, to complete the six different color-resolving tasks in the figure 1, four kinds of color samples are effectively distinguished. Therefore, the multi class identification problem can be transformed into multiple two class identification problems. The two-classification problem is simple, and it can quickly lock the optimal features for the classification and remove the invalid features as much as possible. If each of the two classifications can find the optimal effective features, the feature subsets of these features can be very well identification all faults. According to the above analysis, a feature selection method (TFS) is proposed to ensures that feature subset have excellent distinguishing power for any two kinds of faults.

As a new feature selection strategy, the target feature selection method has the following characteristics: 1) A heuristic search strategy [16]. 2) Feature subset identification performance complements each other, which ensures the global identification ability of feature subsets. 3) Feature identification capability guides feature selection. Feature identification capability evaluation uses a filter-based evaluation criterion [16-18]. 4) High computational efficiency, small number of feature subsets, and good identification.

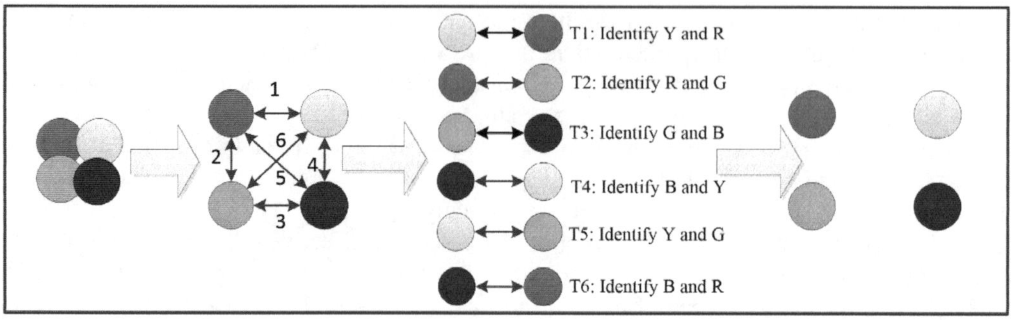

Figure 1. Identification process of multiple color samples

The TFS structure is shown in figure 2. Target setting: decomposing the N item classification into the C_n^2 item two classification task, each of the two classification tasks is called a Target Group (TG). Target feature selection: For each TG, a filter-based feature evaluation criterion is used to find the most effective feature to distinguish the target, and the target's optimal feature is called the Target Feature (TF). Target feature fusion: With a reasonable fusion of the target features of the entire target group, a new subset of features can be obtained, which is an optimized feature subset of the N item classification problem.

Step1: First of all, according to the target structure, for all the fault samples, two different fault samples constitute a target sample space.

Step2: In each target sample space, according to the feature evaluation criteria, all features are evaluated and selected to get the best identification feature of the target, that is, the target feature of the target.

Step3: Each target sample space is performed by step 2, and the target feature of each target are obtained, and then the target features are reasonably fused to form the optimized feature subset.

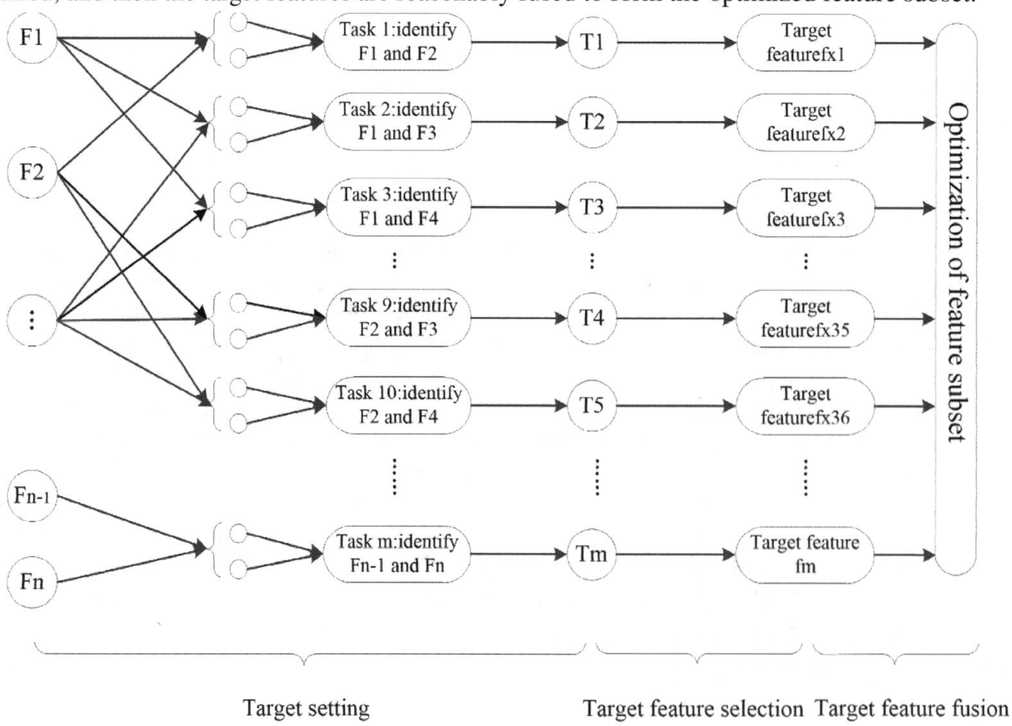

Figure 2. Target feature selection structure

3. Evaluation criteria

In pattern characterization and pattern identification, feature optimal selection is advantageous to pattern characterization and identification, and feature selection depends on feature evaluation. The feature evaluation criterion proposed in this paper mainly considers the dispersion between categories and the compactness within categories [16-18]. However, for feature dispersiveness between fault categories, not all features have good dispersion among the fault categories, and there are also some features of overlapped overlaps between the faults cases. In this case, the fault identification ability of the feature is determined by the number of samples in the mixed region of the fault feature range. The feature evaluation criterion is determined by the degree of fault sample aliasing.

3.1 Single target feature evaluation criteria 1

The feature evaluation criteria are mainly determined by the feature dispersion between the fault categories and the compactness of the features within the category. This paper uses the difference method to describe the compactness of features within a fault category, as Eq1. d_m^i is the compactness of the feature i in the category of fault m ; fy_m^i 、 fx_m^i are the maximum and minimum of the feature i in the fault category m .

$$d_m^i = fy_m^i - fx_m^i \tag{1}$$

The dispersion of features between fault categories is another important factor in the feature evaluation criteria. The boundary difference method is used to measure the dispersion of the feature between the fault categories, as equation (2). $H_{m,n}^i$ is the dispersion of the feature i between fault category m and fault category n .

Combining the compactness and dispersion of feature, the average Euler distance is used to evaluate the feature's fault identification ability [14-21]. The larger the average Euler distance of the feature, the stronger the fault identification capability, as equation (3). $D_{m,n}^i$ is the average Euler distance of the feature i between fault category m and fault category n .

$$H_{m,n}^i = \begin{cases} fx_n^i - fy_m^i, fx_n^i \geq fy_m^i \\ \\ fx_m^i - fy_n^i, fx_m^i \geq fy_n^i \end{cases} \tag{2}$$

$$D_{m,n}^i = \frac{2 \times H_{m,n}^i}{d_m^i + d_n^i} \tag{3}$$

3.2 Single target feature evaluation criteria 2

For the cases where some of features ranges mixed overlap between the faults, the fault identification ability of the feature is determined by the number of samples in the mixed region of the fault feature range. The feature evaluation criterion is determined by the degree of fault sample aliasing, as equation (4). The greater the degree of aliasing of features, the weaker the fault identification capability.

$$P_{m,n}^i = \frac{N_m^i + N_n^i}{2 \times N} \tag{4}$$

$P_{m,n}^i$ is the degree of aliasing of features i between fault category m and fault category n; N is the total number of each category of fault samples; N_m^i is the number of the fault category sample m belongs to the aliased region.

4. Applications in fault diagnosis

The experiment and engineering application based on the above theory are introduced in this section.

The experiment original data includes 9 fault categories, each with 118 samples. The application experiments are conducted according to figure 3. According to TFS, (x_1, x_2, x_6, x_{12}, x_{19}, x_{32}, x_{42}, x_{53}, x_{63}, x_{70}, x_{76}, x_{79}, x_{83} x_{85}, x_{95}, x_{96}) is selected as the feature set to identify the faults. Fault identification ability value of target feature are showed in table 1. The comparison of target features with other features in same TG are listed in figure 4. The results of 9 faults diagnosis are showed in table 2, and the comparison of different methods in table3.These experiments are carried out under the same condition with the same data.

In the same target sample space, compared with other features, the target feature has good distinctiveness and can classify an unknown sample by this distinction., as Figure 4 shown. This confirms that the feature set selected by TFS is much more prominent in fault diagnosis. It can be seen from table 3 that this method not only guarantees higher accuracy of fault diagnosis, but also has certain advantages for multi-category fault diagnosis, compared to several other feature optimization and selection methods.

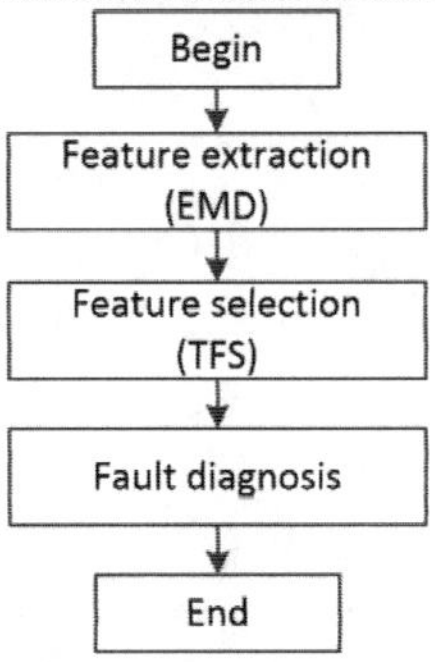

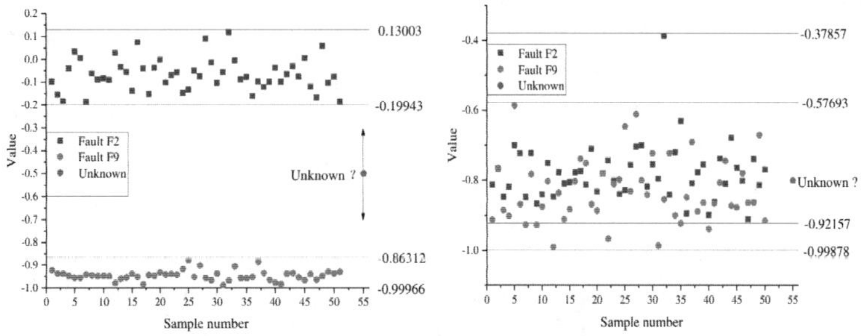

Figure 3. the flow of experiment **Figure 4.** the comparison of features

Table 1. Target feature fault recognition ability

Target group sample	Target feature	Average Euler distance	Aliasing degree	Target group sample	Target feature	Average Euler distance	Aliasing degree
F1，F2	x_6	3.581	--	F3，F7	x_{96}	0.049	--
F1，F3	x_{76}	0.709	--	F3，F8	x_1	-	0.093
F1，F4	x_{53}	7.703	--	F3，F9	x_{76}	1.059	--
F1，F5	x_2	4.504	--	F4，F5	x_{19}	2.081	--
F1，F6	x_{19}	1.550	--	F4，F6	x_{32}	0.955	--
F1，F7	x_{42}	0.247	--	F4，F5	x_{63}	1.183	--
F1，F8	x_{19}	0.715	--	F4，F8	x_{53}	3.475	--
F1，F9	x_1	--	0.076	F4，F9	x_{53}	2.632	--
F2，F3	x_{95}	1.149	--	F5，F6	x_{19}	1.463	--
F2，F4	x_{12}	7.031	--	F5，F7	x_{70}	0.791	--
F2，F5	x_{12}	3.891	--	F5，F8	x_3	1.609	--
F2，F6	x_{12}	0.869	--	F5，F9	x_{32}	1.997	--
F2，F7	x_{83}	1.088	--	F6，F7	x_{32}	0.657	--
F2，F8	x_{83}	2.201	--	F6，F8	x_{32}	1.275	--
F2，F9	x_{83}	3.106	--	F6，F9	x_{32}	1.419	--
F3，F4	x_6	0.883	--	F7，F8	x_{79}	--	0.016
F3，F5	x_{85}	0.632	--	F7，F9	x_{96}	0.224	--

F3, F6	x_{32}	0.562	--	F8, F9	x_{76}	0.397	--

Table 2. The results of fault diagnosis by TFS

Fault category	F1	F2	F3	F4	F5	F6	F7	F8	F9
Diagnostic accuracy	100%	100%	89.1%	100%	100%	100%	90.8%	99.7%	100%

Table3. The comparison of different methods

Feature selection method	Classification	Number of fault	Diagnostic accuracy
PCA	SVM	4	98.75
MPE	ANN	4	96.53%
DET-PSO	PNN	6	97.23%
TFS	**PNN**	**9**	**97.74%**

MPE (Multiscale permutation entropy) [20]; DET-PSO (Distance evaluation technique particle swarm optimization)[12]; LDA (Independent component analysis) [21];

5. Conclusions

A feature selection method based on EMD and TFS is presented and discussed for fault diagnosis in this paper. On the basis of filter evaluation criterion and heuristic search strategy, this novel feature selection method selects the optimal feature for each target sample space. It not only could remove the redundant information to the hilt, but also retains the most obvious difference between samples. Moreover, TFS is easy to design and could complete the calculation with great rapidity. Therefore, TFS effectively improves the accuracy of fault diagnosis, and it is a very useful feature selection method for multi fault diagnosis.

References

[1] Nandi S, Toliyat H A, Li X. Condition Monitoring and Fault Diagnosis of Electrical Motors—A Review[J]. *IEEE Transactions on Energy Conversion*, 2005, 20(4):719-729.

[2] Choi S, Akin B, Rahimian M M, et al. Implementation of a Fault-Diagnosis Algorithm for Induction Machines Based on Advanced Digital-Signal-Processing Techniques[J]. *IEEE Transactions on Industrial Electronics*, 2011, 58(3):937-948.

[3] Zhang W F, Yan H. Exon prediction using empirical mode decomposition and Fourier transform of structural profiles of DNA sequences[J]. *Pattern Recognition*, 2012, 45(3):947-955.

[4] Cherif L H, Debbal S M, Bereksi-Reguig F. Choice of the wavelet analyzing in the phonocardiogram signal analysis using the discrete and the packet wavelet transform[J]. *Expert Systems with Applications*, 2010, 37(2):913-918.

[5] Ali J B, Saidi L, Mouelhi A, et al. Linear feature selection and classification using PNN and SFAM neural networks for a nearly online diagnosis of bearing naturally progressing degradations[J]. *Engineering Applications of Artificial Intelligence*, 2015, 42(C):67-81.

[6] Huang … [et. al. The empirical mode decomposition and the Hilbert spectrum for nonlinear and non-stationary time seri[M]. *The Royal Society*, 1998.

[7] Amirat Y, Benbouzid M E H, Wang T, et al. EEMD-based notch filter for induction machine bearing faults detection[J]. *Applied Acoustics*, 2018, 133:202-209.

[8] Zheng J, Cheng J, Yang Y. Generalized empirical mode decomposition and its applications to rolling element bearing fault diagnosis[J]. *Mechanical Systems & Signal Processing, 2013, 40(1):136-153.*

[9] Piao Y, Park H W, Jin C H, et al. Ensemble method for classification of high-dimensional data[C]// I*nternational Conference on Big Data and Smart Computing*. IEEE, 2014:245-249.
[10] Hu L, Gao W, Zhao K, et al. Feature selection considering two types of feature relevancy and feature interdependency[J]. *Expert Systems with Applications*, 2018, 93.
[11] Tang T, Bo L, Liu X, et al. Variable predictive model class discrimination using novel predictive models and adaptive feature selection for bearing fault identification[J]. *Journal of Sound & Vibration*, 2018, 425:137–148.
[12] Van M, Kang H J. Bearing-fault diagnosis using non-local means algorithm and empirical mode decomposition-based feature extraction and two-stage feature selection[J]. *Science Measurement & Technology Iet*, 2015, 9(6):671-680.
[13] Babu N R, Mohan B J. Fault classification in power systems using EMD and SVM[J]. *Ain Shams Engineering Journal*, 2015, 8(2).
[14] Kumar I, Virmani J, Bhadauria H S, et al. Classification of Breast Density Patterns Using PNN, NFC, and SVM Classifiers[M]// *Soft Computing Based Medical Image Analysis*. 2018.
[15] Tomar D, Agarwal S. Feature Selection based Least Square Twin Support Vector Machine for diagnosis of Heart Disease[J]. *International Journal of Bio-Science and Bio-Technology*, 2014, 6:69-82
[16] Chen X, Zhou J, Xiao J, et al. Fault diagnosis based on dependent feature vector and probability neural network for rolling element bearings[J]. *Applied Mathematics & Computation*, 2014, 247(C):835-847.
[17] Chen X, Zhou J, Xiao H, et al. Fault diagnosis based on comprehensive geometric characteristic and probability neural network[J]. *Applied Mathematics & Computation*, 2014, 230(3):542-554.
[18] K. Fujarewicz, M. Wiench. Selecting differentially expressed genes for colon tumor classification[C]// *Int. J. Appl. Math. Comput. Sci.* 2003:327-335.
[19] Chen X Y, Zhou J Z, Xu X M, et al. A Hierarchical Feature Selection Method Based on Classification Tree for HGU Fault Diagnosis[J]. *Advanced Materials Research*, 2014, 1037:398-403.
[20] Vakharia V, Gupta V K, Kankar P K. A multiscale permutation entropy based approach to select wavelet for fault diagnosis of ball bearings[J]. *Journal of Vibration & Control*, 2014, 21(16).
[21] Widodo A, Yang B S, Han T. Combination of independent component analysis and support vector machines for intelligent faults diagnosis of induction motors[J]. *Expert Systems with Applications*, 2007, 32(2):299-312.

IOP Conf. Series: Materials Science and Engineering **630** (2019) 012024 doi:10.1088/1757-899X/630/1/012024

Fault Diagnosis Based on Tree Heuristic Feature Selection and FS-DFV for Rolling Element Bearings

Xiaoyue chen[1,2*], Xiong Liu[1], Dang Ge[1], and Mengchao Liu[1]

[1] School of Electrical Engineering and Automation, East China Jiaotong University, Nanchang 330013, China

[2] Huazhong University of Science and Technology, Wuhan, Hubei 430074, China

*997921925@qq.com; wdmxdefeng@126.com; 976932632@qq.com; 601075460@qq.com

Abstract. In order to make up for the deficiency of traditional single diagnosis in rolling element bearing fault diagnosis application, eliminate a large amount of redundant information and improve the classification effect of the aliasing mode, based on comprehensive analysis of the respective advantages of fuzzy set and tree search, this paper presents a joint rolling bearing fault diagnosis method based on tree-inspired feature selection and FS-DFV (Fuzzy Set and Dependent Feature Vector). The dependent feature vectors (DFV) can dig deeper the essential differences of the faults and improve the fault accuracy. By establishing the heuristic tree model, the tree type heuristic feature search strategy is designed, and the excellent feature selection criteria based on the density clustering with noise are proposed, and the conventional feature selection model is improved. In addition, fuzzy set are used to process the problem of extracting aliasing patterns in the DFV, and fuzzy membership is used to guide subsequent feature extraction of the alias modes. The proposed method is compared with the other four fault diagnosis methods. The experimental results show that the proposed method can effectively improve the diagnostic efficiency of the rolling bearing.

1. Introduction

In modern industry, rolling element bearing (REB) is one of the most basic components of rotating machinery [1]. It is widely used in aerospace [2], engineering manufacturing [3] and other industries. If the machine is operated for a long time, the rolling bearing part will be worn, peeled off, etc., and as a result, the service life of the machine will be reduced, and even the life of the personnel will be threatened, resulting in irreversible consequences. Therefore, the accurate condition monitoring and fault diagnosis of REB play an important role in ensuring the reliable operation of the machine. Effective diagnosis of the REB has become a hot spot in current research.

For REB, it is currently the most popular strategy to collect bearing vibration signals and perform fault diagnosis. In this way, predecessors have developed many analytical methods, such as envelope spectra and wavelet transforms [4-6]. since the Empirical Mode Decomposition (EDM) method was proposed by Huang [7], the complex signal can be decomposed into simple signals in dealing with non-stationary and nonlinear data. Because of this feature, it is widely used in signal processing, and its advantages are more obvious than traditional signal processing [8-10].

The use of EDM to extract feature signals of rolling bearings does not effectively remove a large number of redundant features, so feature parsing has become the mainstream of optimization algorithms. Combining multi-scale entropy and least squares support vector machine, the rolling bearing signal can be optimized to form a eigenvector containing only the main information and achieve significant improvement [11].By combining the fast bearing empirical mode decomposition and the stochastic decrement technology bearing fault detection method, fault features can be quickly extracted and fault types analyzed [12].The real-valued gravitational search algorithm (RGSA) is used to optimize the input weights and biases of ELM, and GSA is used to select important features from the composite feature set to achieve good results [13].Among them, the dependent feature vector (DFV) proposed by Chen Xiaoyue was applied to the fault diagnosis of REB [14], and its unique characteristic organizational form opened up new diagnostic methods. At the same time, with the continuous optimization of intelligent algorithms, support vector machines, artificial neural networks and other classifiers are widely used in the fault diagnosis of rolling bearings [15-17], and have achieved good results. However, there are still the following problems: 1) Boundary effects and aliasing modes still exist. 2) The classification choice is too complicated.

Aiming at the above situation, this paper designs a tree heuristic feature search strategy, introduces fuzzy set to process aliased regions, and constructs fuzzy set and dependent vector (FS-DFV). DFV, in their unique organizational form, can fully exploit the characterization capabilities of each feature, and at the same time shield the interference of redundant information on classification. The addition of fuzzy set improves the efficiency and self-healing of fault diagnosis. Therefore, FS-DFV constructed by combining the advantages of the two can overcome the problem of poor discrimination in the aliasing area in fault diagnosis. Based on the improved intelligent optimization algorithm proposed in this paper, the fault diagnosis efficiency is improved, and the timeliness and practicality of the algorithm are increased.

2. Feature evaluation

Whether or not the heuristic tree can select the most effective features, feature evaluation plays a decisive role. Therefore, this paper adopts two aspects of single feature evaluation and feature set evaluation to comprehensively evaluate, so as to ensure the high efficiency of selected features in fault diagnosis.

2.1. The evaluation of one feature

In pattern characterization and pattern recognition, the selected features must be able to distinguish different faults. At the same time, the compactness of the fault data is very good under this feature. in order to achieve these indicators, this paper proposes a density-based clustering evaluation method. density-based spatial clustering of applications with noise (DBSCAN) can be divided into any cluster, and noise can be filtered out. The figure 1 is based on density clustering and fault clustering comparison.

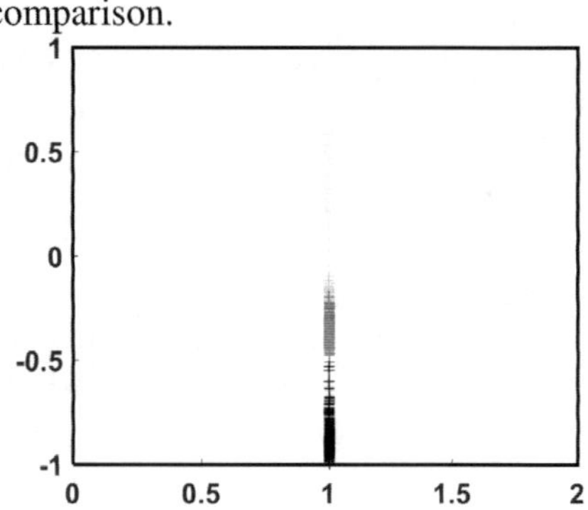

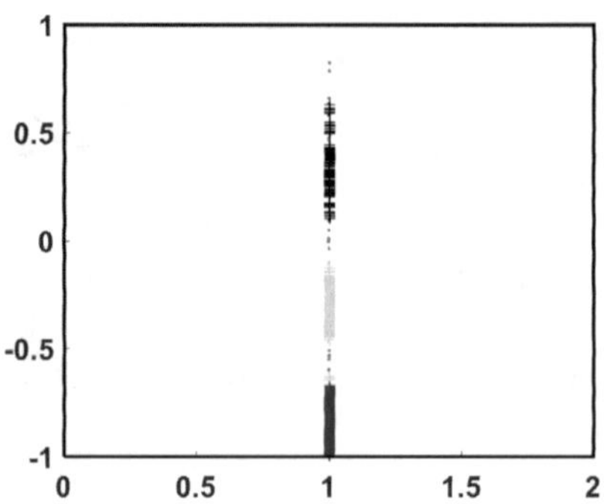

Figure 1. Density clustering scatter diagram

2.2. The evaluation of a feature set

The smaller the Euler distance in the same sample class, the better the sample size is better, the greater the Euler distance between the different sample classes, which indicates that the two samples based on this feature have a large distance interval, which can be very good for fault diagnosis. Euler distance is often used to measure the advantages and disadvantages of feature subset in pattern recognition.

$$d_{ij} = \frac{1}{h \times k} \sum_{m-1}^{h} \sum_{n-1}^{k} MSE(i(m), j(n)) \tag{1}$$

Formula: MSE is Euler distance, i(m) and j(n) are two data sets respectively, h and k correspond to the number of data contained in these two datasets respectively. Dij is an average Eulerian distance.

3. search strategy

3.1. feature selection

Based on the above feature evaluation, the most effective feature is selected and the complex sample space is divided into only one type of sample, and the features used are recorded. The optimized feature subset is formed, and the flow chart of fuzzy dependent feature selection is showed in figure 2.

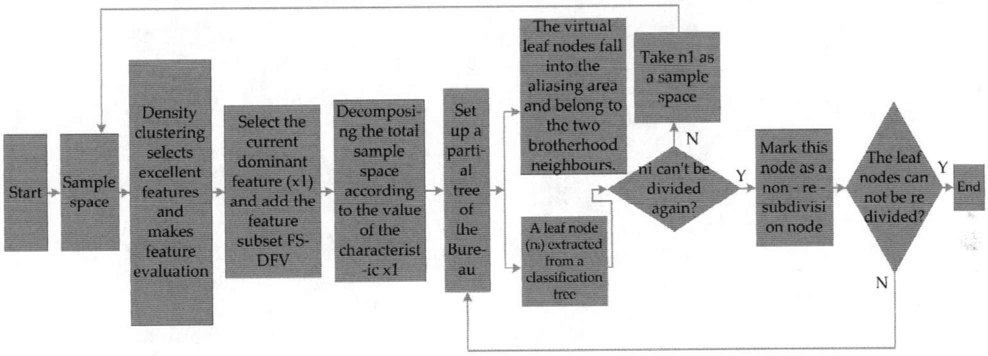

Figure 2. Flowchart of hierarchical fuzzy feature selection method based on classification tree

3.2. Feature extraction based on fuzzy set

In order to deal with the aliasing mode, this paper introduces fuzzy sets. The most important part of fuzzy set is the construction of fuzzy membership degree. There is no specific form of fuzzy membership function. Variance is often used to describe whether discrete variable distribution is compact. As shown in equation (2), Di represents the *i*th subspace. The compactness of the fault on feature x, the smaller its value indicates better compactness.

$$D_i = \frac{1}{n}[(x_1 - \overline{x_i})^2 + (x_2 - \overline{x_i})^2 + \ldots + (x_n - \overline{x_i})^2] \tag{2}$$

Where n represents the number of feature x in one of the fault samples in the subspace; x_n represents the nth eigenvalue of a fault sample; xi represents the mean of the eigenvalues of the *i*th fault sample in the subspace.

According to the structure of the tree search, the sample mean and variance of each fault in the parent sample space of the aliased region are selected as a mean sample set and a variance sample set, and the equation (3),(4) is as follows. When judging for each aliased region, the corresponding variance and mean of each fault sample are different.

$$\overline{X_i} = \{\overline{x_1}\ \overline{x_2}\ \ldots\ \overline{x_n}\} \tag{3}$$

$$D_i = \{d_1\ d_2\ \ldots\ d_n\} \tag{4}$$

The difference between the eigenvalues scattered in the fuzzy region and each fault in the fault space under this feature is used to make the difference. The absolute value of the result is worse than the above, and the membership function the equation 5 is as follows:

$$\mu_A(x_i) = e^{-\frac{|x_j - \overline{X_i}|}{D_i}} \tag{5}$$

Where x_j is the *jth* fault point that falls in the fuzzy region, and $\overline{X_i}$ is the average value of the *ith* fault sample characteristic; D_i is the variance of the fault type to which the *ith* fault point belongs;

IOP Conf. Series: Materials Science and Engineering **630** (2019) 012024 doi:10.1088/1757-899X/630/1/012024

When establishing a FS-DFV each virtual node is set as a fuzzy set correction part (FSCS) to determine which kind of fault each aliased area characteristic value belongs to because the sample mean and variance in the membership function are a data set. Therefore, in each judgment, the obtained fuzzy membership degree value is also a set of numerical values, and the largest value is selected to determine the basis for assigning the value of the aliased region. The equation 6 is as follows, FS-DFV Extraction tree combining fuzzy set with DFV is showed in figure 3.

$$\mu_A(x) = \max(\mu_A(x_1)\ \mu_A(x_2)\ ...\ \mu_A(x_n)) \tag{6}$$

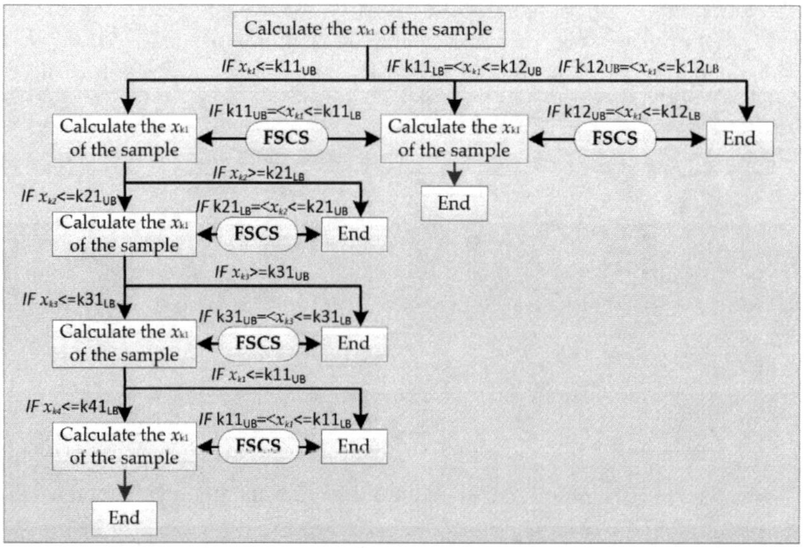

Figure 3. FS-DFV extraction tree combining fuzzy set with DFV

4. Fault diagnosis

This section describes experimental and engineering applications based on the above theory. Figure 4 shows the fault diagnosis flow. Figure 5 shows the comparison of the DFV and FS-DFV extraction rates. Figure 6 shows the three-dimensional view of the failure types before and after the FS-DVF treatment. Table 1 shows the diagnostic rates of DFV and FS-DFV on a probabilistic neural network, and table 2 shows the comparison of five vectors for fault diagnosis.

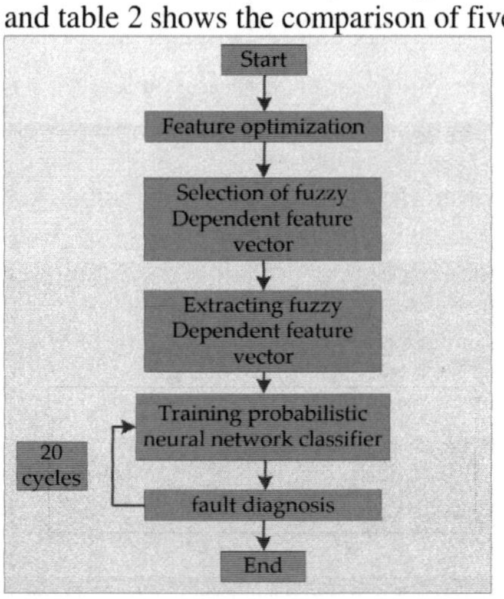

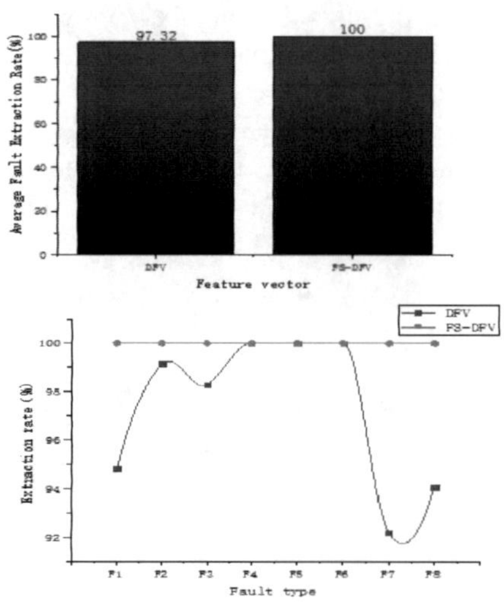

Figure 4. Flow chart of fault diagnosis **Figure 5.** The comparison of the DFV and FS-DFV

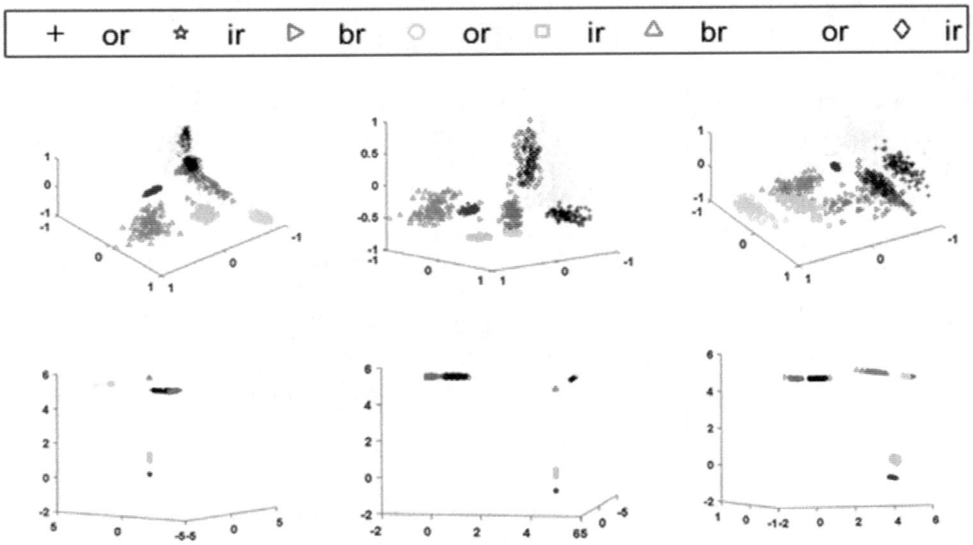

Figure 6. Comparison of the FS-DFV before and after processing

Table 1. FS-DFV fault diagnosis results

Fault classes	FS-DFV the diagnostic rate	DFV the diagnostic rate
FF_1	100%	91.32%
FF_2	100%	99.04%
FF_3	97.57%	95.44%
FF_4	100%	100%
FF_5	100%	100%
FF_6	100%	100%
FF_7	96.18%	90 .49%
FF_8	97.21%	93.38%

Table 2. Comparison of fault diagnosis of rolling bearing

Feature vector	The average diagnostic rate	The feature extraction time (ms)	The training time (ms)	The testing time (ms)
FS-DFV	**98.87%**	**114**	**377**	**254**
DFV	96.62%	109	369	258
SSF	95.26%	219	380	249
FSR	71.24%	221	393	246
AF	96.71%	2793	1435	1243

In table 2, by comparing the other four methods, the average diagnostic efficiency of the FS-DFV fault diagnosis method is better than the other four methods, thus verifying the efficiency of this method. The corresponding time is also within a reasonable range. The time used for the diagnosis method based on FS-DFV is far less than that of the original feature. This shows that the FS-DFV can be used to represent the fault samples to obtain a satisfactory diagnosis efficiency. By comparing with the DFV, the edge blurring process can improve the insufficiency of the DFV and improve the accuracy.

5. Conclusions

In this paper, a fault diagnosis algorithm for REB is proposed, which combines the FS-DFV and tree search. In the fault diagnosis, a fuzzy set is introduced. The fuzzy membership function can be used to deal with the edge problem effectively. Combining tree search and density clustering, the key features of the fault are screened to determine the structure of the FS-DFV, and the validity of the fuzzy membership is verified by the extraction of the FS-DFV of the original data. The experiment shows that the method is high in the extraction of FS-DFV. Efficiency. By the probability neural network, the five of feature vector fault diagnosis is compared. The FS-DFV should be compared with the other four of feature vector in the fault diagnosis efficiency because of the effect of the other four of feature vector, and the time it takes in the ideal range. The comprehensive fault diagnosis method based on FS-DFV and tree type heuristic feature selection has both breakthrough accuracy and excellent computational efficiency. It is an excellent fault diagnosis method.

References

[1] He G, Ding K, and Lin H 2016 Fault feature extraction of rolling element bearings using sparse representation *Journal of Sound & Vibration* p 514-527

[2] Silaev B M, Fedorov D G, and Danilenko P A 2017 Dynamic of Damage of Aircraft Engine High-speed Rolling-element Bearings Tested in Low-viscosity Model Mediums *Procedia Engineering* p 43-49

[3] Cheng Y, Zhou N and Zhang W 2018 Application of an improved minimum entropy deconvolution method for railway rolling element bearing fault diagnosis *Journal of Sound & Vibration* p 53–69

[4] Singh J, Darpe A K, and Singh S P 2018 Rolling element bearing fault diagnosis based on Over-Complete rational dilation wavelet transform and auto-correlation of analytic energy operator *Mechanical Systems & Signal Processing* p 662-693

[5] Tyagi S Panigrahi S K 2017 An improved envelope detection method using Particle Swarm Optimisation for rolling element bearing fault diagnosis *Journal of Computational Design & Engineering*

[6] Klausen A, Robbersmyr K G, and Karimi H R 2017 Autonomous Bearing Fault Diagnosis Method based on Envelope Spectrum p 13378-13383

[7] Huang N E, Shen Z, Long S R, et al. 1998 The Empirical Mode Decomposition and the Hilbert Spectrum for Nonlinear and Non-Stationary Time Series Analysis *Proceedings Mathematical Physical & Engineering Sciences* p 903-995

[8] Sungho Park, Seokgoo Kim, and Joo-Ho Choi 2018 Gear fault diagnosis using transmission error and ensemble empirical mode decomposition *Mechanical Systems and Signal Processing* p 262-275

[9] Bigdeli S, Danandeh H, and Ebrahimi M M 2017 A correlation based bullet identification method using empirical mode decomposition *Forensic Science International* p 278:351

[10] Rai A Upadhyay S H 2017 Bearing performance degradation assessment based on a combination of empirical mode decomposition and k-medoids clustering *Mechanical Systems & Signal Processing* p 16-29

[11] Li Y, Zhang W, Xiong Q, et al 2017 A rolling bearing fault diagnosis strategy based on improved multiscale permutation entropy and least squares SVM *Journal of Mechanical Science & Technology* p 2711-2722

[12] Xiang J Zhong Y 2017 A fault detection strategy using the enhancement ensemble empirical mode decomposition and random decrement technique *Microelectronics Reliability* p 317-326

[13] Luo M, Li C, Zhang X, et al 2016 Compound feature selection and parameter optimization of ELM for fault diagnosis of rolling element bearings *Isa Transactions* p 556-566

[14] Chen X, Zhou J, Xiao J, et al 2014 Fault diagnosis based on dependent feature vector and probability neural network for rolling element bearings *Applied Mathematics & Computation* p 835-847

[15] Liu H, Zhou J, Zheng Y, et al 2018 Fault diagnosis of rolling bearings with recurrent neural network-based autoencoders *Isa Transactions* p 221-230
[16] Patel J P Upadhyay S H 2016 Comparison between Artificial Neural Network and Support Vector Method for a Fault Diagnostics in Rolling Element Bearings *Procedia Engineering* p 390-397
[17] Srinivas R S, Tiwari R, and Kannababu C 2018 Application of active magnetic bearings in flexible rotordynamic systems – A state-of-the-art review *Mechanical Systems & Signal Processing* p 537-572

An ontology-driven approach for modelling TMS fuel consumption information subsystem

V Kurganov[1], M Gryaznov[2], A Dorofeev[3,4]

[1]Tver State University, 33 Zhelyabova Street, Tver, 170100, Russia
[2]Nosov Magnitogorsk State Technical University, 38 Lenin Street, Magnitogorsk, 455000, Russia
[3]Financial University under the Government of the Russian Federation, 49 Leningradskiy Avenue., Moscow, 125167, Russia
[4]National Research University Higher School of Economics, 33 Kirpichnaya Street, Moscow, 105187. Russia

adorofeev@hse.ru

Abstract. The use of information technologies for the automotive transport and logistics management has opened game-changing opportunities for business, providing solutions to critical problems of sustainable development in turbulent environments. Today, the data for transport operations can be processed and used for the managerial decision-making, actually, on-line. However, despite abundance of TMSs on the market, many transport companies use them for solving operating management problems only. There are also lots of companies which do not use TMSs at all. Practice shows that managers of numerous transport companies have a little idea of how the records management data can be transformed into the awareness of improving efficiency of the transportation process. In our opinion it may be explained by the fact that different management levels use the knowledge from various domains, such as transport, IT, economic, and financial areas. Accordingly, the transformation of contemporaneous data into the managerial decisions requires an approach which is discussed in this paper using an example of introduction of TMS fuel consumption information subsystem.

1. Introduction

Motor transport is one of the most important economic sectors providing operation of many supply chains. Most of the production and commercial facilities, public industries and organizations depend on stable, efficient and uninterrupted operation of these chains. Today's information technologies offer exciting opportunities to transform business processes of both large and small enterprises, including transport and logistics companies. Nowadays, multiple transport management information systems commonly named as Transportation Management Systems (TMS) are present today both on the Russian and foreign markets. It should not go unnoticed that large companies use special modules of ERP enterprise systems for transport management. It has been affirmed that TMSs as a separate class of information systems generated historically from the SCM solutions; and SCM, in turn, generated from the ERP systems [1, 2]. It is well known that ERP systems continued the MRP and MRP II concepts, which had been created for the effective industrial production management. And now these large ERP systems, such as the SAP or Oracle, include transportation process management subsystems [3].

Thus, according to this chronological review we can see that the electronic document management and the supporting IT solutions were created initially for the information support of material flows, and later for the support of financial flows. Accordingly, the data of the electronic workflow generally described operations with goods and money. Now, under conditions of extreme competition in the commodity and trade markets, and the development of e-commerce featuring a glut in the commodity market, the role of transportation has significantly increased, because the customer's requirements for timely delivery, reliability and appropriate costs has also increased [4, 5].

The special transport and cargo monitoring systems based on GPS/GLONASS/RFID technologies have become widely used with the development of satellite-based object tracking facilities and wireless telecommunications. They provided an opportunity to obtain real-time data about the location of vehicles, fuel consumption, vehicle failures and driver behaviour [6, 7]. And it became the impetus for the development of ad hoc subsystems for fleet management - Fleet Management System. And generally, as follows from the literature review [8], the transportation management system (TMS) should include solutions for a broad spectrum of tasks.

2. The most important tasks of transportation management

We have questioned 3,712 Russian companies, and the questionnaire showed that the most of respondents gave priority to fuel recording systems. However, it should be noted that Excel spreadsheet tables are very popular, despite the widespread acceptance of transport management software solutions developed by 1C, Russian company (see figure 1). Whereby, the percent of companies having small stock of vehicles (10 to 50 vehicles) and using various IT solutions is higher than for large companies. And the owners for mixed fleets of vehicles, including not only trucks and passenger cars, but also road construction machinery, prevail among the large companies.

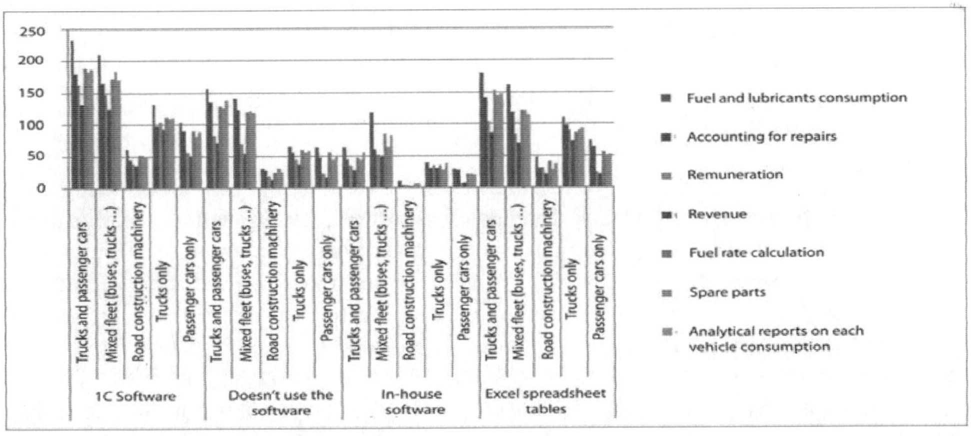

Figure 1. Use of information applications in various transport companies

Thus, the variety of vehicle types at the fleets generates the diversity of data. And in the current context, we should note that the diversity of data, its accessibility and speed of receiving, in turn, have generated the problem of transformation of these data into information required for the managerial decision-making and knowledge. Accordingly, the large number of studies reveals the gap between real needs of business and current TMSs, which is number of studies reveals the gap between real needs of business and current TMSs, which is particularly indicative for large companies. It takes place because such abundance of various information systems providing records management has no analytics and decision support. This, in turn, leads to the inefficient investments into the information technologies on the part of business operating the road transport, as well to the disappointed expectations and disillusions for the owners of these companies [9, 10, 11]. In this regard, the transport and logistic companies as well as other enterprises do not hurry to invest into the development of their actual TMSs and to set about for introduction of innovative IT-solutions.

However, it should be noted that the decisions for selection of one or another TMS are accepted on the basis of both functional and non-functional criteria which include the reputation of IT solutions on the market, total cost of ownership, system flexibility, etc. An organizational structure of a company as well as the body of employees knowledge about the management for transportation process shall play the important role [12]. And, as practice shows, the successful introduction depends generally on how the transport business is ready to consider TMS as the assets which can generate additional value for a transport company [13]. Whereon the owners of cargo-carrying companies have seen and realized that the TMS provides the business transparency and transportation reliability control, we can observe the unfailing interest to the further data extraction using the daily business transactions fixing the transport operations and knowledge which are necessary for important strategic decision making [14]. Thus, the introduction of IT solutions leads to the transformation of transportation business providing the competitiveness through the quick and effective solutions using the data from information systems [15, 16].

3. Modelling and development of the fuel consumption subsystem

Unfortunately, practice demonstrates a lot of cases when the transport companies don't see their benefits in the introduction of TMS; they don't realize the opportunities they create for managerial decision-making at operations, tactic and strategic level, as well as the additional value that can be created for the transportation business. This is especially true for small and medium-size transport and logistic companies both for Russia and other countries of the world [17]. Usually, the large companies have the experience of implementing ERP systems; they understand the approaches to IT project implementation in general terms. However, the success of TMS deployment within a company often depends on the efforts of various specialists with high qualifications and solid hands-on experience. It is worth noting that these experts should not have the similar level of interest in deploying TMS. It is not unusual when their interests do not overlap or are directed differently.

For example, when we consider the task of fuel metering, we can distinguish the following actors It is known that the fuel consumption of a truck depends on a host of factors, such as specifications of the vehicle itself, payload, driving speed, road and weather conditions, tyre pressure [18], as well as on driving skills. [19]. Consequently, the analysis of these factors impact on the transportation process, transport business as a whole as well as their mutual influence requires their hierarchical description and determining the semantic connections between them, all in order to provide the creation of knowledge necessary for managerial decision-making. Thus, we come to the need of forming an ontological model for fuel consumption. It is well-known that an ontology is described using the following data set:

$$O = < T_0, R, P > \tag{1}$$

wherein
T_0 is the set of domain terms;
R is the semantically valuable relationships;
P is the defining functions for interpretation.

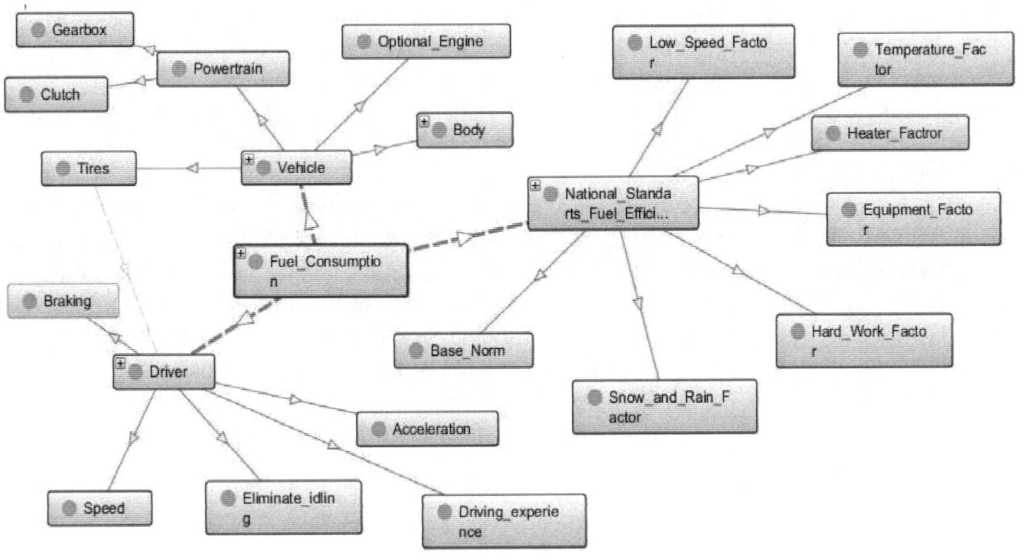

Figure 2. Ontological model of fuel consumption

A literature review shows that driving style causes up to 30% of fuel consumption [20]. In [21], the following key driving factors that increase fuel consumption are highlighted

- engine performance at increased RPMs;,
- frequent speedups and brakings;
- tickover;
- suboptimal route, in terms of the road traffic;
- poor skills of driver during the maneuvering in tough road conditions;

Based on truck design, we conclude that fuel consumption is impacted by follow-ing factors: type, power and other characteristics of the engine; parameters of clutch and gearbox; tyre pressure and tyres themselves; number of center shafts; aerodynamic qualities, etc. Besides, in Russia as well as in many other countries there are standards of fuel efficiency with the basic rate dependent on the make and model of vehicle. The multiplying coefficients exist, too; they depend on the road and weather conditions, traffic in the city, tough driving conditions, age of vehicle, etc. The fuel consumption depends on the payload and the dead weight of the semitrailer.

Thus, we have built the ontological model of fuel consumption (see figure 2). The ontological model allows to evaluate the relationships between the objects associated with the fuel consumption in TMS. As a result, for example, the owner of the company understands that the design characteristics and the age of vehicle have significant impact on the fuel consumption; he would make balanced decisions concerning the renewal of fleet and selection of vehicles. An extra attention should be paid to the selection and further training of drivers as well. The dispatcher should take a more responsible approach at the operating level to the selection of a driver for every individual cruise, taking into account the route, weather and climatic conditions and road traffic. The driver, for example, should be ordered to make a periodic inspection of tyre pressure in case an automated monitoring device is not available.

4. Conclusion

The development and deployment of information systems for transport management require all the stakeholders at various levels to adhere to a common interpretation of the fundamental concepts as well as the cause and effect relations between them. Eventually, the developing architecture of TMS should not only meet the current requirements of business objectives, but also enable to jump-start the business of a specific transport company. Despite the fact that our ontological method requires closer cooperation and trust from all concerned parties, it allows the effective use of knowledge and expertise of various specialists. Thus, it creates the prerequisites for successful application not only of the technological and financial capital, but the social one as well. In the current context of information economy, often called the "knowledge economy", the role of people's competencies, human relations and cooperation increases dramatically. In this regard, the example discussed above shows that relying on the cooperation between a transportation business employees creates not only the ability to formalize and optimize business processes, but also ensure their engagement and motivation. As you know, the question why certain management events and operations happen is a key to development of any enterprise architecture. The proposed method was applied successfully in a number of large and medium-sized Russian businesses, and proved its effectiveness.

5. References

[1] Scavarda L F, de Carvalho A B and Vieira M S 2006 A Reference Matrix for Information System in Supply Chain Management *Brazilian Journal of Operations & Production Management* vol 3(1) pp. 21-48

[2] Verwijmeren M 2004 Software component architecture in supply chain management *Computers in Industry* vol 53 pp. 165–178

[3] Daithankar J and Pandit T 2014 *Transportation Management with SAP TM 9. A Hands-on Guide to Configuring, Implementing, and Optimizing SAP TM* (Apress) p. 328

[4] Holcomb M C and Manrodt K B 2015 24th annual study of logistics and transportation trends: serving up friendly freight. *Logistics Management* vol 9 pp. 26-32

[5] Schulz J D 2018 Big or small, speed and service rules: Smaller, niche freight payment companies with emphasis on shipper service and nimble technology are holding fast with industry giants *Logistics Management* vol 1 pp 28-30

[6] Tyszko A, Templin T and Oszczak S 2007 GNSS systems in vehicle fleet management. *Transport Problems* vol 2(2) pp. 11–16

[7] Saghaei H 2016 Design and Implementation of a Fleet Management System Using Novel GPS/GLONASS Tracker and Web-Based Software *1st International Conference on New Research Achievements in Electrical and Computer Engineering* (IEEE:Tehran).

[8] Reinhold S, Laesser C and Bazzi D 2015 The intellectual structure of transportation management research: A review of the literature *Working Paper Series 005* (St.Gallen :University of St. Gallen)

[9] Dorofeev A 2013 Problems of informatization of transport and logistics companies. *Logistica* vol 4 pp. 26-28

[10] Kurganov V and Dorofeev A 2017 On the practical approach to the implementation of sustainable development of transport and logistics enterprises using information technology *Prospects for the development of logistics and supply chain management* (Moscow: SCM Consulting) pp. 279-297

[11] McCrea B 2012 5 trends driving TMS growth: top supply chain software analysts break down the drivers that will keep the transportation management systems market on a steady upswing for years to come *Logistics Management* vol 1 pp 42-44

[12] Wycoff D F 2009 *Implementing an Sap Transportation Management System Solution: a Case Study* (Regis: Regis University).

[13[McCrea B 2014 2014 state of TMS: cost reductions and ROI continue to soar: Top supply chain software analysts assess growth in the transportation management systems (TMS) market, highlight emerging trends in adoption, and predict the future of this highly beneficial, yet consistently underused application. *Logistics Management* vol 1 pp. 38-42

[14] Hoffman M 2012 Transportation Management Systems: an array of affordable solutions offers fleet managers better visibility and higher profitability *Food Logistics* vol 1-2 pp. 37-39

[15] Holcomb M C, Liao-Troth S and Manrodt K B 2014 A shift in fundamentals: the changing direction in logistics and transportation management *Transportation Journal* vol 53(4) pp. 516-533

[16] Kurganov V and Dorofeev A 2016 The Practice of Business and IT Integration in the Transport Company Using Enterprise Architecture Framework. *Proceedings of 7th International Conference on Operations and Supply Chain Management* (Phuket: Mahidol University) pp. 377-392

[17] Ji Y and Cai L 2012 *Transportation management system in China. A case study on UPS and JULC* (Gävle: University of Gävle)

[18] Sandberg, T 2001 *Heavy Truck Modeling for Fuel Consumption Simulations and Measurements.* (Linkoping; Linkoping University)

[19] Deierlein B 2001 Managing fuel consumption: while the driver is the single most important influence on fuel consumption, in fact there can be as much as a 35-percent difference between the 'best' and 'worst' drivers; good specs can help a fleet improve its fuel economy. *Fleet Equipment* vol 12 pp. 24-27

[20] Carrese S, Gemma A and La Spada S. 2013 Impacts of driving behaviours, slope and vehicle load factor on bus fuel consumption and emissions: a real case study in the city of Rome *Procedia - Social and Behavioral Sciences* vol 87, pp. 211–221

[21] Ferreira J C, Almeida J and Rodrigues da Silva A 2015 The Impact of driving styles on fuel consumption: a data warehouse and data mining based discovery process *IEEE Transactions on Intelligent Transportation Systems* vol 16 (5) pp. 2653-2662

Pattern recognition in the differentiated image for the powder and granulated materials particle size classification

D V Yunovidov[1,2], M N Nadezhin[1,2] and V A Shabalov[2]

[1] JSC "NIUIF", 162622, Severnoe shosse 75, Cherepovets, Russia
[2] Cherepovets State University, 162622, office 203, Lunacharsky str. 5, Cherepovets, Russia

Dm.Yunovidov@gmail.com

Abstract. The paper shows and investigates the technique of classifying the particle size of powder and granulated materials. The objects of the study are industrially produced mineral fertilizers. Samples with different composition (about five types of mineral fertilizers) and various degrees of particle size (less than 100 μm, less than 500 μm and granules of 2-5 mm) were examined. The samples particles have an irregular shape, close to spherical (in the case of granules) or cubic (in the case of powders). To improve the accuracy and eliminate the particle shape influence on analysis, the preliminary pressing of samples on a boric acid substrate was used. The keynote of the proposed technique is to obtain an optoelectronic image of an object with a resolution of at least 640x480 pixels (a three-dimensional matrix of pixel intensity in the Red-Green-Blue (RGB) system). Next, the area of analysis is separated from the obtained image and transformed into grayscale (a two-dimensional matrix of pixel intensities with a resolution of at least 200x200 pixels). The influence of external illumination (gradient, temperature and brightness) is eliminated by the grayscale image differentiation. The result is the "surface map" of the sample, which reflects defects in the pressed structure (patterns, which are responsible for the size of the particles). According to the found patterns, the samples are classified according to their particle size. Four classification algorithms were investigated (linear, linear with L1 and L2 regularization, and a nonlinear "random forest"). All proposed approaches are automated and implemented in the Python 3.6 programming language. There is provided the selection of the operating parameters of all the described algorithms.

1. Introduction

Many authors write about a new industrial revolution in recent years (Industry 4.0) [1,2]. This phenomenon is characterized by the use of flexible and decentralized industrial processes. It also applies to the quality control of products. The phenomenon of Industry 4.0 and flexible control are particularly relevant for large and actively modernizing facilities, where the control is based on visual indicators. Examples of such industries are metallurgy [3], cement industry [4,5] and mineral fertilizers production [6,7]. One of important quality parameters in these industries is the particle size distribution of granular and powder materials [4,8,9]. With this parameter the production of mineral fertilizers is reviewed (where the particle size is responsible for the quality of the technological process [6,7], the quality of pellets [10,11], the agronomic effect [12] and the quality of sample preparation for chemical and physical analyses [13]). Methods of sieve analyses [8,13], laser light scattering (LLS) [4,14] and optoelectronic control [4,9,15,16] are often used to determine the particle size. Unfortunately, they all have a number of disadvantages. The sieve analysis cannot provide continuous control of the particle size distribution and it highly depends on the shape of the particles [8]. The LLS is a fundamental and highly accurate method of analysis. However, it is limited by the size of the studied particles and cannot estimate the shape and the colour of materials [4,14]. The third group of methods (optoelectronic control) is the most universal. It is based on work with the image of particles and allow us to estimate a wide range of parameters. Optoelectronic control is widely used in industrial production [5,9]. There are several examples where this method provides control of the shape and colour of particles [15,16]. However, the optoelectronic method has a significant drawback - it strongly depends on external measurement conditions of illumination (gradient, brightness and temperature). The solution of this problem isn't really covered in the studied literature.

The purpose of this work is to develop an automated method for the classification of granular and powder materials by its particle size using the optoelectronic image with elimination of dependence on the external illumination (brightness, temperature, and gradient).

2. Objects and Methods

2.1. Objects of the analysis and sample preparation
In this work there are 5 types of mineral fertilizers investigated (products of the JSC "Apatit"). Each sample was prepared as raw granules (2-5 mm) and grounded powder with two fractions (< 500 μm and < 100 μm) in 10 series. Further, samples were pressed into tablets in the form of a "sandwich structure" (in molds of 20 mm in diameter). At first, the plate of boric acid (about 3 g) was made by pressing with a figure punch and 100 bar force (3 t/cm^2) for 30 seconds. Then the mineral fertilizer sample (about 1.5 g) was put into the boric acid plate and pressed with a usual punch at 260 - 280 bar (10-11 t/cm^2) for 60 seconds. About 150 samples were prepared.

2.2. Obtaining the optoelectronic signal
Pressed samples were placed on a white sheet of paper and were photographed at a fixed distance to the sample surface (10.0 ± 0.5 cm) with natural light illumination (sufficient for visual determination of the surface structure of the sample). A digital USB video camera was used as an optoelectronic device for obtaining the image (frame resolution of at least 640x480 pixels, focal length 2.8 - 12 mm and a sensor of ½ 7`` CMOS). Furthermore, the image (figure 1) was transferred to a computer via USB protocol. For each image we can observe difference in brightness, temperature, and gradient of illumination.

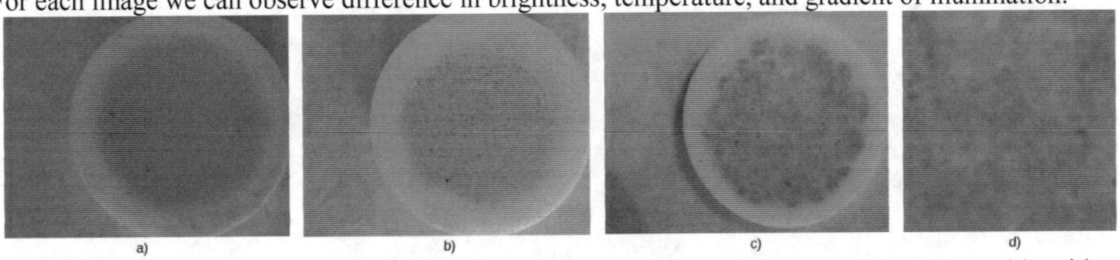

Figure 1. The example of the objects images. The three different fertilizers with <100 μm (a); with <500 μm (b) and with 2-5 mm granules (c) and the 200x200 pixels area of the analysis (d).

2.3. The image processing algorithm
In this paper, the authors solve the novel problem of optical analysis of powder size with varying external illumination (brightness, temperature, and gradient). The possibility of classifying samples using the set of standard algorithms is evaluated. The work uses the following approach.

- Record of the objects image with a resolution of at least 640 × 480 in RGB format.
- Select the surface area (at least 200 x 200 pixels) and calculate it average brightness (feature No.1).
- Transform image into a grayscale (two-dimensional matrix of pixel brightness intensities).
- Differentiate the image to eliminate the effect of illumination. The sample "surface map" is formed as the result (the two-dimensional array, which characterizing the pressed sample surface).

- Smooth the "surface map" by the median filter (it preserves the boundaries of the patterns [17]).
- Calculate the average number and average area of patterns on the "surface map" (relative to image area) by the "marching square" algorithm [18] (feature No.2 and 3).
- Classify the particle size according to the value of found normalized features.

All stages were selected experimentally for the studied objects, providing standardization of external conditions and work with sufficient accuracy. Adjustable parameters of image processing algorithms are the smoothing window for median filter and "contour constant" for the "marching squares" algorithm [19]. Further, with obtained features (the average number and area of the patterns and the average brightness of the image) the "objects-features" matrix for classification is compiled. Using a compiled matrix and the information about particle size of the pressed material, the different classification approaches were calculated (linear without regularization, linear with L1 and L2 regularization, and non-linear algorithm of "random forest"). The F-measure (harmonic mean of precision and recall) was chosen as the quality metric [20]. The classification algorithms were tested using a cross-validation strategy (with 30% of randomly selected objects from the total set with preserving the distribution of target classes). All proposed algorithms were implemented in the Python 3.6 programming language.

3. Results and discussion

According to the described method, about 150 samples of 5 types of fertilizers were analysed. For each image of the sample, a "surface map" was made (figure 2).

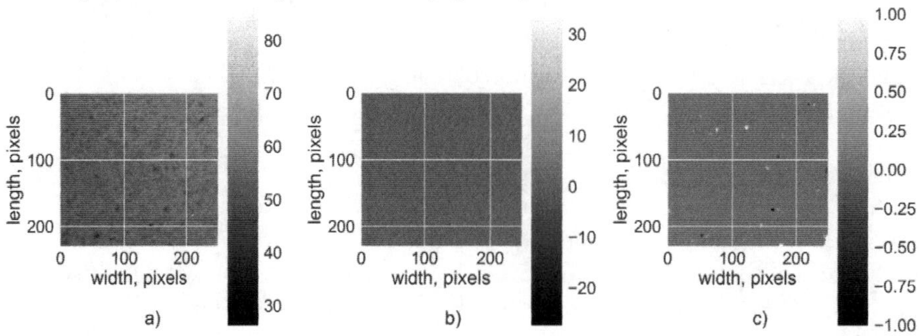

Figure 2. The original (a); differentiated (b) and smoothed (c) "surface map" of pressed powder.

Further, the patterns (pixels in "surface map" whose intensity is different from 0) were approximated as closed contours. Their average area and quantity were calculated using (1) and (2) formulas.

$$N_r = \frac{N}{a \cdot b} \tag{1}$$

Where N_r is the average number of patterns; N is the number of found patterns; a and b are the length and width of the image in pixels.

$$S_r = \frac{\sum_{i=1}^{N} S_i}{a \cdot b} \tag{2}$$

Where S_r is the average area of the patterns; S_i is the area of the i pattern; N is the number of patterns; a and b are the length and width of the image in pixels.

As the addition feature we have calculated the average brightness of the image in grayscale (3).

$$I_g = \frac{\sum_{i=1}^{N} I_i}{N} \times I_w \tag{3}$$

Where I_g is the average brightness of the image surface; I_i is the brightness of the i pixel (from 0 to 255 rel. units); N is the number of pixels; I_w is the average brightness of the white background. It was established experimentally that the multiplication with average white intensity (not the division or subtraction) made the finest contribution to the particle size classification.

Parameters of smoothing the image (the range from 1 to 31 with a pace of 5 pixels) and the approximation of patterns (the range from 0.1 to 2.0 with a pace of 0.1 relative units) were selected from the range to improve the further classification. We have used the method of the "grid search". Spearman correlation between the average amount (and average area) of the calculated patterns and the size of the sample particles was used as the quality metric (figure 3).

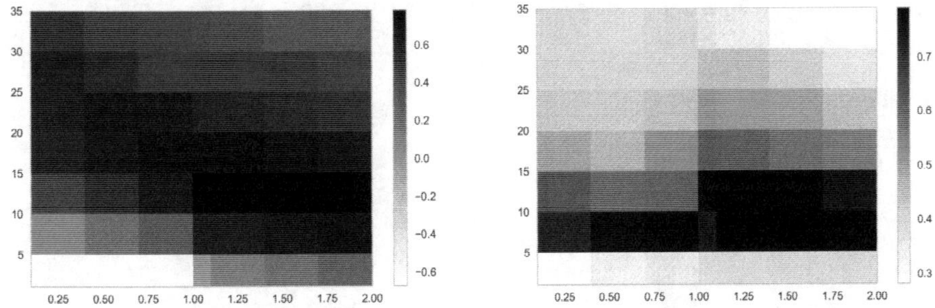

a) b)

Figure 3. Map of changes in the Spearman correlation between the particles size and the average amount (a) or average area (b) of the patterns.

The optimal parameters were 13 pixels and 1.3 units. It is interesting to note that the Spearman correlation also works well in the range of parameters 5 – 15 pixels and 0.5 – 1 units. In this regard, a calculation of the number of patterns was made (figure 4).

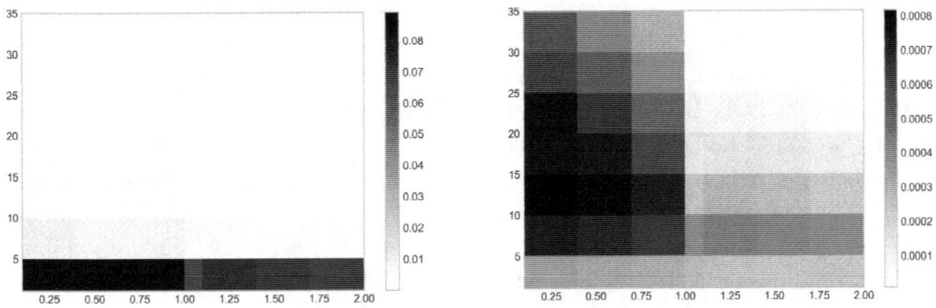

a) b)

Figure 4. Map of changes in the number of average quantity (a) and average area (b) of patterns.

With previously selected parameters of 15 pixels and 1.3 units, the values of the number and area of patterns tend to 0. This is caused by poor recognition of patterns on samples with small particles size (< 100 μm). At the same time, an increase in the correlation coefficient is caused by a reduction in the number of classes. Thus, the contour constant should not exceed 1 and the best values of the coefficients were 7 pixels and 0.8 relative units.

With obtained parameters and features the several classification algorithms were investigated. The target variable is the particle size in the pressed sample (categorical variable, with values < 100 μm, < 500 μm and 2-5 mm). We also select configurable parameters of each classifiers [21] using the grid search. To study the effect of selected features on the classification, their significance were also calculated (4) (table 1).

$$im_j = \frac{w_j}{\sum_1^n w_i} \times 100 \quad (4)$$

Where the im_j is the "importance" of the j feature in %; w_j is the weight of the j feature in the classification equation, w_i is the weight of the i feature in the classification equation.

The random forest algorithm shows the best results (93%). This behaviour is predictable. The number and area of patterns of the image are related to each other and do not necessarily linear depend on the particle size. However the linear algorithms show the good results too (86%). The most significant feature were the average number of defined patterns. Actually, for pressed samples the variability of the defects area will be small. Consequently, their weight for linear classification will also be small. On the contrary, for the non-linear algorithm the significance of features is slightly equalized. Separately, we note that the average brightness of pixels plays a smaller role for the classification. This suggests that the differential correction of the illuminate works well.

Table 1. Classification results and significance of features.

	Linear regression with L2 regularization (Ridge)	Linear regression with L1 regularization (Lasso)	Linear regression	Random forest
	Achieved F-measure, %			
Average	83	86	86	93
Standard deviation	4	6	6	3
	Feature significance, %			
Average number	56	64	59	43
Average area	30	21	23	37
Average brightness	14	15	18	20

4. Conclusion

As the result of this work, the novel automated method for the classification of granular and powder materials by its particle size and correction method for the illumination in optoelectronic image (gradient, temperature and brightness) were proposed. The classification of the particle size according to the selected features of average number and area of image patterns was carried out based on various industrially produced mineral fertilizers. The work of 4 different classification algorithms (both linear and non-linear) was investigated. The quality of the classification was achieved in 93 % by F-measure after the parameters were chosen. The best result was shown by the "random forest" algorithm (the number of "trees" is 82, the maximum number of signs is unlimited, the maximum depth of each tree is unlimited, the bootstrap is false, weighed registers are true). The best linear algorithm is the linear classification with or without L1 regularization (86% by F-measure). The achieved quality of the classification indicates that the proposed technique can be used in industrial control. The proposed approaches are automated and quite simple to implement. Experimental data is provided [22].

5. References

[1] Meissner H, Ilsen R and Aurich J 2017 Procedia CIRP 62 165-9
[2] Qin J, Liu Y and Grosvenor R 2016 Procedia CIRP 52 173–8
[3] Murphy T and Schade C 2019 Additive Manufacturing for the Aerospace Industry 99 142
[4] Lilkov V, Dimitrova E and Gaidardzhiev S 1999 Cem. Concr. Res. 29(1) 3–8
[5] Pasquazzi A, Weissenbacher R and Oehlers M 2017 Int. J. Refract. Met. Hard. Mater. 62 118–25
[6] Swisher D, Borgelt S and Sudduth K 2002 Trans. ASAE 45(4) 881–8
[7] Zhang R, Wang X, Guo J, Chen L, Zhou J and Ma W 2014 Sensors & Transducers 26 1–6
[8] Besler H. 2008 Dev Sedimentol. 59 73–98
[9] Bjørk T, Mair K and Austrheim H 2009 J. Struct. Geol. 31(7) 637–53
[10] Wiącek J, Molenda M and Stasiak M 2018 Particuology 39 88–95
[11] Yohannes B, Liu X, Yacobian G and Cuitiño A 2018 Adv. Powder Technol. 29(12) 2978–86
[12] Tarragó E, Puig S, Ruscalleda M, Balaguer M and Colprim J 2016 Chem. Eng. J. 302 819–27
[13] Kimura M 2013 Testing Methods for Fertilizers (Tokio: Incorporated Administrative Agency Food and Agricultural Materials Inspection Center) p 370
[14] 2009 International Standard ISO 13320 First Edition 2009-11 (United Kingdom)
[15] Chávez GM, Sarocchi D, Santana E and Borselli L 2015 Comput. Geosci. 85 248–57
[16] 2006 International Standard ISO 13322-2 First Edition 2006-11 (United Kingdom)
[17] Strugailo V 2012 Sci. Educ. Bauman MSTU 12(5) 270–81
[18] Mantz H, Jacobs K and Mecke K 2008 J. Stat. Mech. Theory Exp. 12 2-28
[19] Scikit-image Module 2019 [Online] Available: http://scikit-image.org/docs/dev/api/skimage.measure.html#skimage.measure.find_contours
[20] Powers D 2011 J. Mach. Learn Technol. 2(1) 37–63
[21] Scikit-learn Module 2019 [Online] Available: https://scikit-learn.org/stable/supervised_learning.html
[22] Yunovidov D 2019 mineral fertilizer images (pressed samples) Mendeley Data, v1 [Online] Available: http://dx.doi.org/10.17632/2yw4bnbz5m.1

Authors' background

Your Name	Title	Research Field	Personal website
Dmitry Yunovidov	Ph.D, lecture	Big data analysis, Analytical chemistry, Automation, Optical recognition, Mineral fertilizer industry	https://www.dimyun.space/
Maksim Nadejin	master student	Big data analysis, Automation, Optical recognition	-
Viktor Shabalov	Ph.D, lecture, associate professor	Automation, Optical recognition	-

Analysis of Orbital Movement Lunar Orbital Station

Du Chongrui[1, 2], Vsevolod V. Koryanov[1], Chen Danhe[2]

[1]Bauman Moscow State Technical University, Russia
[2]Nanjing University of Science and Technology, China

ducrui@163.com, vkoryanov@mail.ru, cranefeeling@hotmail.com

Abstract. Represents the new project "Deep Space Gateway (DSG)" - an orbital station in the vicinity of the Moon [1]. The DSG can be used as a staging post for exploration missions to the lunar surface and eventually to other deep space destinations including Mars. It is also a platform in a location where the human and technological challenges of long duration human missions in deep space can be investigated and addressed [2]. A DSG may be placed farther from the Moon in halo orbit relative to its position on the line of the Moon-Earth, known as the L_2 Lagrange point. One of the modern orbits of interest as a destination for a spacecraft in cis-lunar space is the Near Rectilinear Halo Orbits (NRHO) [4]. This work is devoted to the study of issues related to the orbital movement DSG, to understand which orbit will be optimal.

1. Introduction

The International Space Station (ISS) was a valuable asset for testing the necessary research and technology for more remote flight scenarios. However, the ISS requires constant replenishment and maintenance and is designed to work exclusively in low Earth orbit (LEO) conditions. Consequently, for the next stage of a manned space flight, intelligence capabilities in addition to the capabilities of the ISS are needed.

Since the return of the Apollo 17 spacecraft to Earth in December 1972, not a single person has gone beyond the limits of low Earth orbit. As a springboard for such research, the focus is currently on developing the ability to support the crew in orbit around the Moon [3].

ESA is seeking information from the European research community to inform about the development of the Deep Space Gateway — the spacecraft in the vicinity of the Moon, which will host crewed missions and operate without crew in between. The Deep Space Gateway is planned to be built and operated during the 2020's as humanities next step beyond Low Earth Orbit and out into the Solar System. Deep Space Gateway is created as a strategic platform, from which human exploration of the Solar System can set forth. Its location in the lunar vicinity, and outside the deep gravitational borehole of the Earth, allows it to be used as an intermediate post for exploratory expeditions to the lunar surface and, ultimately, to other distant space directions, including Mars. It is also a platform where human and technological problems of long-term human missions in deep space can be explored and solved.

The platform is being prepared in the framework of international cooperation under the leadership of partner agencies of the international space station: ESA, NASA, JAXA and CSA [1, 7].

The DSG gets its name because its position is beyond the deepest part of the Earth's gravity. This means that you need less energy to start a mission from there. This makes it an excellent starting point for sending human expeditions to the lunar surface and more distant places such as Mars, i.e. for training for the future flight to Mars. It can also act as a receiving facility for the initial examination (quarantine for the purposes of planetary defence) of samples returned from Mars or other planets.

2. Orbital motion Deep Space Gateway

Deep Space Gateway will be placed in special points in space. For example, where the gravitational pull between the Earth and the Moon is balanced. In some cases, the DSG may be placed farther from the Moon in halo orbit relative to its position on the line of the Moon-Earth, known as the L_2 Lagrange point. The balance of gravitational forces here allows to "park" the spacecraft to make observations.

Table 1. Orbit Types considered for Deep Space Gateway.

Orbit Type
Near Rectilinear Halo Orbits, NRHO
Distant Retrograde Orbit, DRO
L_2 Halo
Low lunar Orbit, LLO

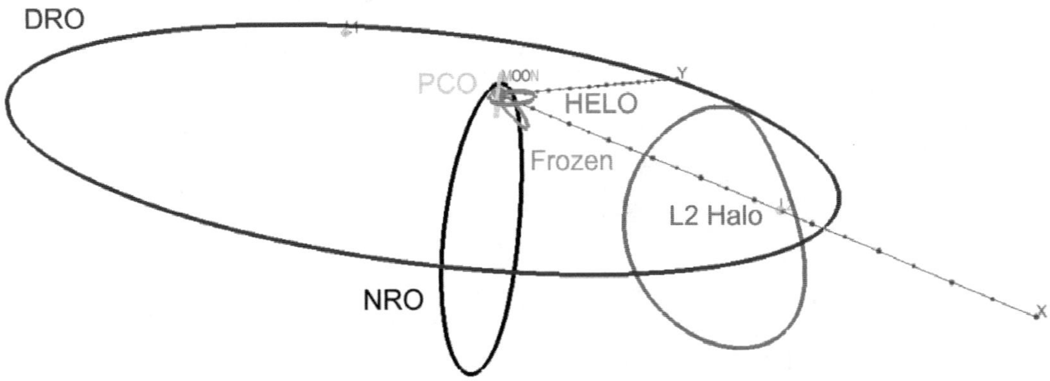

Figure 1. Orbit Types considered for Deep Space Gateway.

One of the modern orbits of interest as a destination for a spacecraft in habitat in the cislunar space, is the Rectilinear Halo Orbits (NRHO). A subset of the halo orbit family, NRHO, is attractive as intermediate orbits for several reasons, including advantageous movements from the Earth and to destinations beyond the Earth's proximity, communication lines, the possibility of limiting eclipse time and favorable access to the lunar surface [4].

This type of trajectory was first identified in a simplified representation of gravitational effects in the Earth-Moon system, that is, in the Circular Restricted Three Body Problem (CR3BP). The study of the motion of the apparatus around the L_2 point will be carried out in the barycenter system of the Earth, Moon bodies.

2.1. Circular Restricted Three Body Problem (CR3BP)

In the model of CR3BP, NRHO are characterized by favorable stability properties, which imply the ability to support NRHO-like movement for a long time, while consuming few fuel resources.

The dynamic model in CR3BP serves as a reasonable approximation to dynamic models with higher fidelity in the Earth-Moon system, including those that may also include solar gravity. As part of this application, the CR3BP considers the motion of a massless spacecraft under the gravitational influence of the Earth and the Moon. It is assumed that these two primary bodies, modeled as point masses, move in circular orbits around their common barycenter. The motion of the spacecraft is then described relative to the coordinate system$\hat{x} - \hat{y} - \hat{z}$, which rotates with the movement of the Earth and the Moon, as shown in Fig. 1[5]. In this frame, the spacecraft is located in dimensionless coordinates (x, y, z). By convention, the values in CR3BP are dimensionless, so the distance between the Earth and the Moon, as well as the mean motion of the primaries, are both equal to a constant

IOP Conf. Series: Materials Science and Engineering **630** (2019) 012027 doi:10.1088/1757-899X/630/1/012027

value of unity. In addition, the masses of the Earth and the Moon are dimensionless and equal to 1-μ and μ, respectively, where the parameter μ is equal to the ratio of the mass of the Moon to the total mass of the system. In a rotating frame of reference, the scalar equations of motion for a spacecraft are written as:

$$\ddot{x} - 2\dot{y} = \frac{\partial U}{\partial x}, \ddot{y} + 2\dot{x} = \frac{\partial U}{\partial y}, \ddot{z} = \frac{\partial U}{\partial z}. \quad (1)$$

где $U = \frac{1}{2}(x^2+y^2) + \frac{1-\mu}{d} + \frac{\mu}{r}$, $d = \sqrt{(x+\mu)^2 + y^2 + z^2}$,
$r = \sqrt{(x-1+\mu)^2 + y^2 + z^2}$.

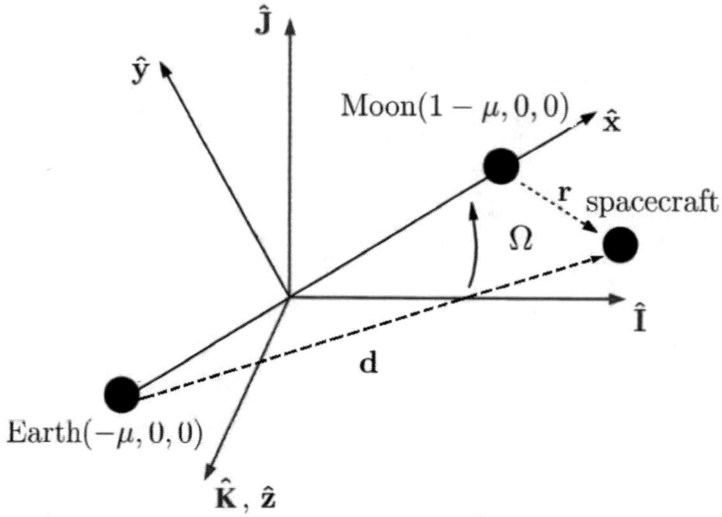

Figure 2. Geometry in CR3BP for the Earth-Moon system

2.2. Some NRHO Qualities and Rating Resources Available in Deep Space Gateway

The creation of a viable intermediate orbit in the cis-lunar space is a key step in the human journey beyond the limits of a low Earth orbit. Both in terms of access from the Earth, access to other destinations, and influencing the design of a spacecraft is all important. Although more work will be done to better understand the properties of cis-lunar orbits, the near rectilinear orbit (NRHO) seems to be the most favorable orbit to fulfil many competing, constraints and requirements[1].

Some relevant properties of NRHO in table 2

Table 2. Some qualities of two conditional orbits for DSG.

Orbit	Property	Value
Near Rectilinear Halo Orbits, NRHO	Period	6-8 days
	Orbit around	Moon
	Distance to the surface of Moon	Approx. From 2000 to 75 000 km
	Inclination	Approx. 90 °
	Visibility from Earth	Constant
Earth-Moon L_2 halo orbit	Period	8-14 days
	Orbit around	Libration point L_2 Earth-Moon
	Distance to Moon	60000 km
	Visibility from Earth	Constant

IOP Conf. Series: Materials Science and Engineering **630** (2019) 012027 doi:10.1088/1757-899X/630/1/012027

3. Near Rectilinear Halo Orbits

This type of trajectory was first identified in a simplified representation of gravitational effects in the Earth-Moon system, that is, in the Circular Restricted Three Body Problem (CR3BP) problem. The study of the motion of the apparatus around the L_2 point will be carried out in the Earth-Moon barycenter system.

The NRHO, in the vicinity of the Moon, potential long-term orbits for a ship with a crew in the wing zone are determined [4].

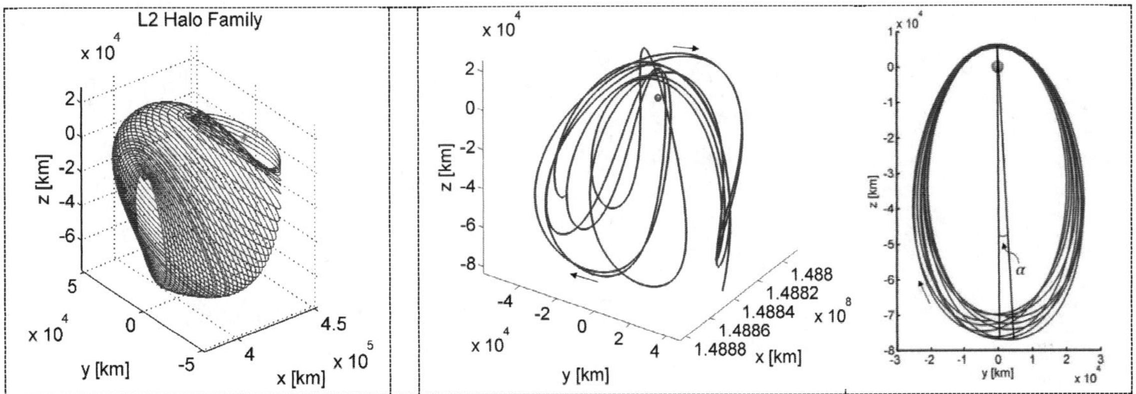

a. In the Earth-Moon rotating system. **b.** In the selenocentric reference system. **c.** In the Earth-Moon rotating system

Figure 3. Near Rectilinear Halo Orbits

3.1. Orbit characteristics

NRHOs have such advantages as relatively inexpensive transfer options from Earth, possible transfer options to the lunar surface and other orbits in cis-lunar space and beyond, and advantageous eclipse properties. Including part of the large L_2 halo-orbit family depicted in Figure 3, the NRHO subset may be roughly limited to specific locations reflecting changes in linear stability across the L2 halo family. [4].

A family of related periodic orbits bifurcates, in fact, from the L_2 near rectilinear halo orbits; members of this family appear in Figure 3(a, b). In appearance, members of this family are a "butterfly" in shape; in particular, this family of orbits fords from a specific NRHO, which has a period of approximately 6 days. The butterfly's orbit family has characteristics similar to those NRHO, offering similar advantages for designing a trajectory in this neighborhood. The orbital movement of the butterfly resembles the shape of the figure eight, but these orbits wrap around the near and far sides of the Moon, as can be seen in Figure 3(c). Figure 3(c). The family of orbits of southern butterflies when viewed in configuration space [5].

4. Orbital analysis

The near-rectilinear halo-orbit was discovered as a kind of bridge between the halo L1 and L_2 in the Earth-Moon system. Since then, many studies have studied the use of the Near-Linear Halo-Orbit for use in various concepts of the exploration of the Moon, including the constant coverage of the South Pole. NRHOs are halo orbits with large amplitudes over the north or South Pole with shorter periods that pass close to the opposite pole. Although they look like large elliptical orbits, they are CR3BP orbits that remain relatively stationary in the Earth-Moon plane, rotating at the same speed as the Moon around the Earth and the Moon around its own axis [3, 8].

If we specifically place the satellite at point L_2 without speed, it will rotate together with point L_2. The white trajectory in fig. is the trajectory of the satellite, which is fixed relative to the point L_2 with the influence of the Sun. It is obvious that no more than one turn, it will fly to the sun.

The numerical solution of the orbital motion is carried out in special software.

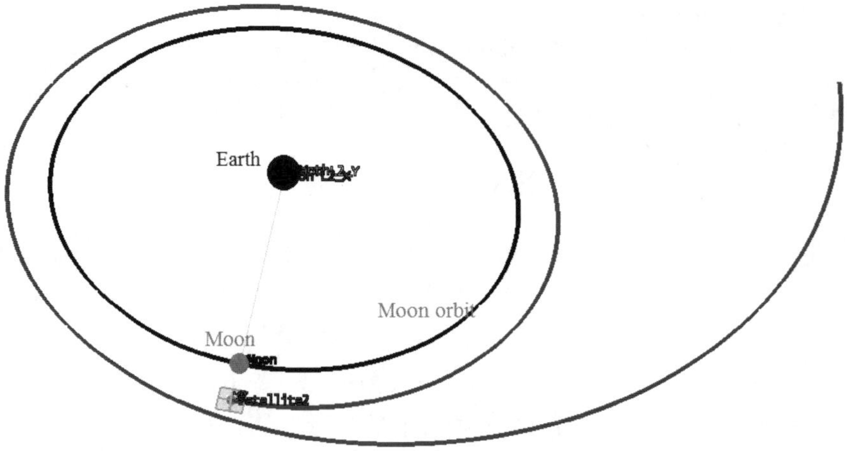

Figure 4. Orbit at libration point

In the system of central gravity of the Earth (Fig. 4) it is seen that the satellite rotates around the Earth with a period of 28 days, it will fly no more than one turn, this means that the trajectory is high unstable.

4.1. Analysis of Halo orbit at libration point

If the satellite is not stationary, but rotates around the point L_2, then the trajectory is halo-orbit, the period is 8-14 days. In the central gravity system of the Moon, it can be seen that the satellite simultaneously rotates around the Moon, and around the L_2 point in the Earth-Moon rotating system, also around the Earth, in the central gravity system of the Earth (in fig.5).

The numerical solution of the orbital motion is carried out in special software.

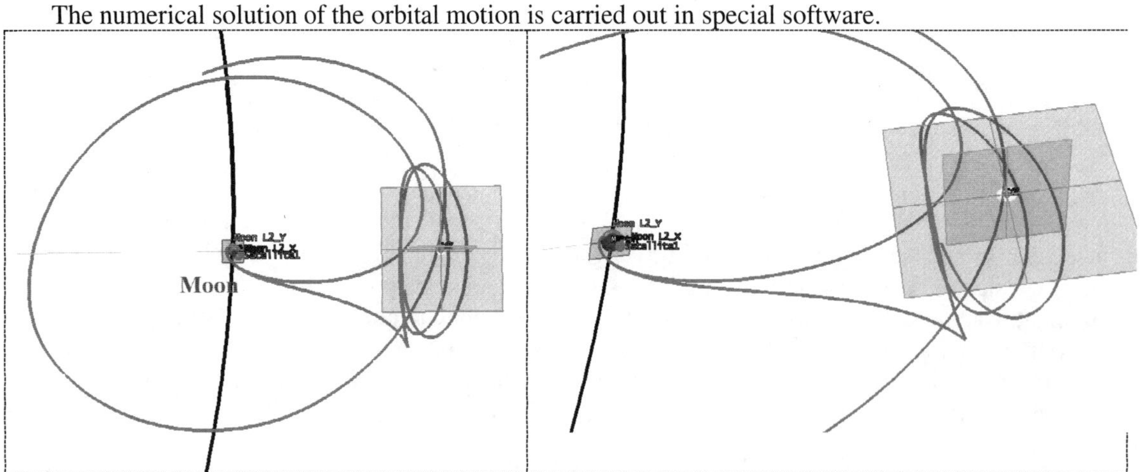

a	b

Figure 5. Halo orbit at libration point

The halo orbit is slightly more stable than the fixed satellite at the point L_2, but even the initial velocity changes fractionally, the halo orbit does not work. See in fig. 6a, 6b.

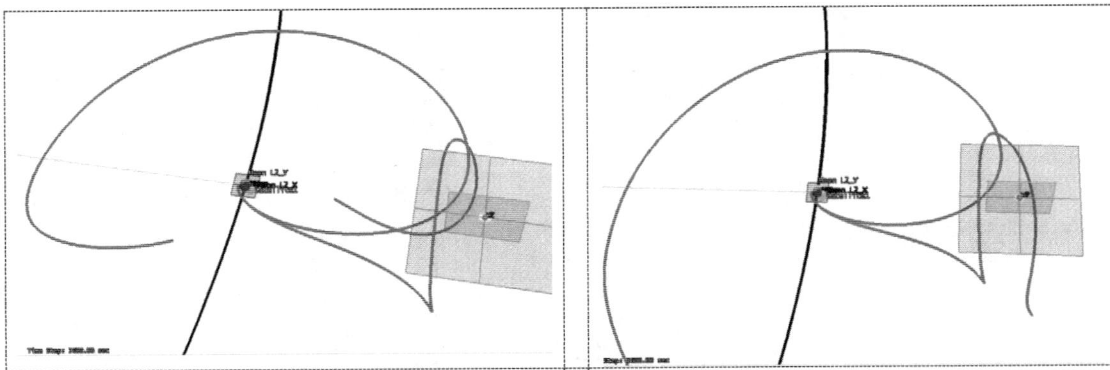

a. With initial speed ΔVx = 0.6059 km / sec

b. With initial speed ΔVx=0.60591km/sec

Figure 6. Final orbit at different incremental speeds

4.2. Analysis of Near Rectilinear Halo Orbits

Influenced by the Moon, the Earth and the sun, it is unstable. From the point of view of the Moon, the DSG repeatedly passes one pole, providing great opportunities for scientific measurements.

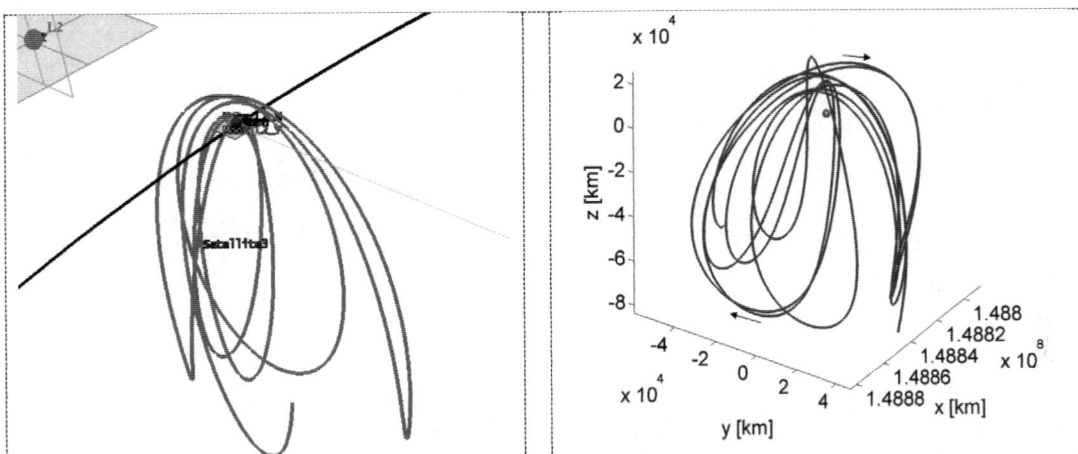

a

b

Figure 7. Near Rectilinear Halo Orbits in the selenocentric reference system

Table 3. The initial parameters of the NRHO in the transfer system of the Cxyz system from the point of barycenter to the point L_2.

Coordinates	Value
X	67000 km
Y	0 km
Z	2200 km
Vx	2 km/s
Vy	0,6 km/s
Vz	0 km/s

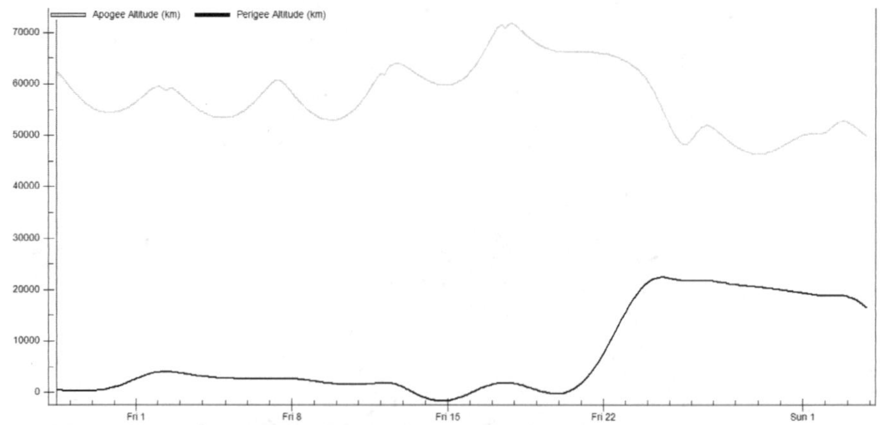

Figure 8. Dependences of the height of perilune and apolune on time

As can be seen in Fig. 8, the height of the apolune is 55000km ÷ 70000km, the height of perilune is almost 2000km ÷ 3000 km. After some time they will lead away if without correction.

In the Earth-Moon rotary system, the trajectory is a more stable ellipse. This means that the spacecraft is always oriented toward the Earth.

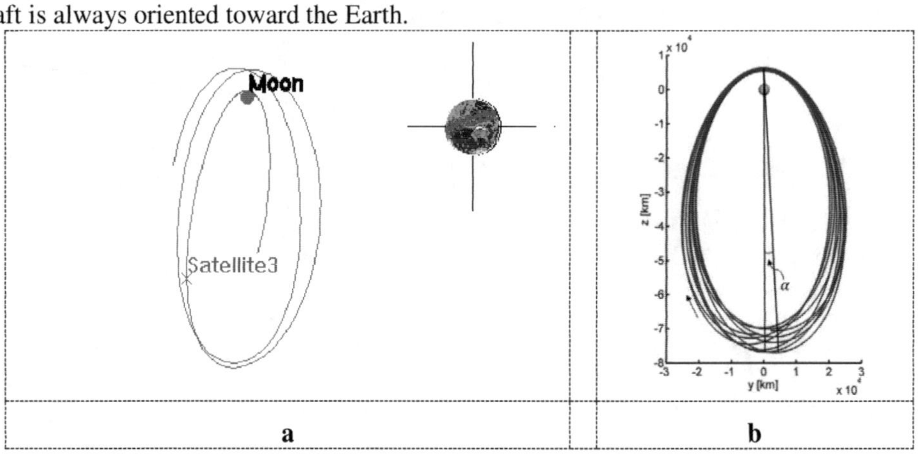

Figure 9. Near Rectilinear Halo Orbits In the Earth-Moon rotating system

In the system of central gravity of the Earth in Fig.10, it is apparently symmetrical with a period of 6-8 days. NRHO with a period of 6.83 days will have a period equal to 1/4 of the lunar sidereal day. Thus, the geometry of the orbit would repeat every lunar orbit. This can make predictable rendezvous graphics more achievable.

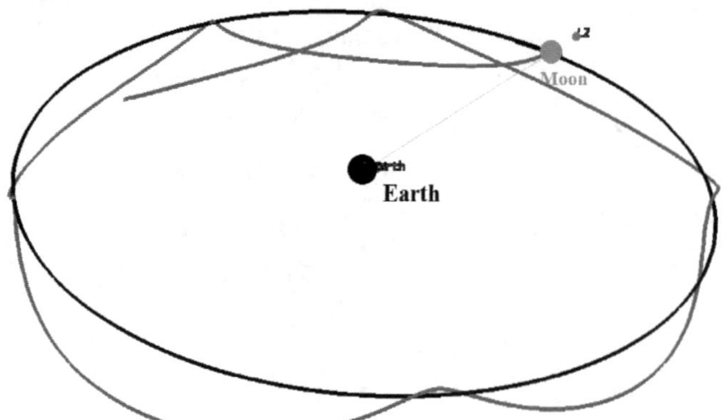

Figure 10. Near Rectilinear Halo Orbits in the system of central gravity of the Earth

5. Conclusion

Studies of the orbit show that these orbits are only quasistable, therefore, to maintain the DSG in these orbits will require some adjustments to the DSG position. The orbits to be used for the DSG mission have advantages and disadvantages. Namely:

The halo orbit at L_2 in the Earth-Moon system shows a good gain in the velocity increment ΔV. However, the remoteness of this orbit from the surface of the Moon is its disadvantage, since it will take a lot of time to fly from this orbit on the surface of the Moon.

For the NRHO orbit, its closer position to the Moon can be distinguished, the flight time is reduced accordingly, but its configuration makes it necessary to constantly recalculate the position of the spacecraft in orbit.

As a result, the main conclusion can be made: the DSG mission provides a great scientific interest, opens up opportunities for exploring the Moon and other planets of the solar system.

The orbital motion for DSG is rather complicated, and it is necessary to study this issue in detail!

References

[1] Deep Space Gateway Research Call Issue 1p1 release - *ESA Workshop: Research Opportunities on the Deep Space Gateway*. Prepared by James Carpenter. Reference ESA-HSO-K-AR-0000. Issue/Revision 1.1. Date of Issue: 27/07/2017.

[2] Deep Space Gateway - lunar station for training for the future flight to Mars. *Latest Space News*. Anatoly Alizar. March 29, 2017.

[3] Martinez, Roland, and Ryan Whitley. "*Options for Staging Orbits in Cis-Lunar Space*." (2016).

[4] Guzzetti, Davide, et al. "Stationkeeping analysis for spacecraft in lunar near rectilinear halo orbits." *27th AAS/AIAA Space Flight Mechanics Meeting*. 2017.

[5] Zimovan, Emily M., Kathleen C. Howell, and Diane C. Davis. "Near rectilinear halo orbits and their application in cis-lunar space." *3rd IAA Conference on Dynamics and Control of Space Systems*, Moscow, Russia. 2017.

[6] Guzzetti, Davide, et al. "Stationkeeping analysis for spacecraft in lunar near rectilinear halo orbits." *27th AAS/AIAA Space Flight Mechanics Meeting*. 2017.

[7] Vsevolod Koryanov, Da-Poian Victoria. Deep space gateway: the future platform for space exploration // *XLII Academic Conference on cosmonautics*, pp. 232-233, 2018.

[8] Vsevolod Koryanov, Da-Poian Victoria. Space servicing: Strategy and logistics // *Proceedings of the International Astronautical Congress,* IAC, 2018-October

IOP Conf. Series: Materials Science and Engineering **630** (2019) 012028 doi:10.1088/1757-899X/630/1/012028

Construction of Modern University Library Website (CMULW) Based on User Needs

Kai Liao, Feng Li

Library, Huazhong Agricultural University, Wuhan, China

309861649@qq.com

Abstract. The construction of modern university library website (CMULW) adapts to the changes of the times, and it will extend the scope and content of library services, which is of great significance to the development of universities. This article will be presented to describes the CMULW must follow the user needs from three aspects, including basic function needs, safety and security needs and humanistic respect needs. The website safety and security needs will be used in the website security system to introduce. The CMULW conforms to the requirements of the current era of social information technology development and provides users with better network information services. The modern university library website (MULW) needs not only to enhance its own practicability, and it needs to satisfy the user omni-directional demand to improve the user experience. Only constantly to find and meet the needs of users and improve service capability and the quality of service, is the basic attitude and development path of the modern university library website construction (MULWC).

1. Introduction

The university library is well stocked, but the collection development of university libraries will be limited by library space. However, the MULW can alleviate the contradiction to some extent. The MULW can integrate library collection resources, set up the digital library, expand the traditional library space, extend the traditional library service mode. It is the basic requirement of MULWC in all-round and high efficiency to meet user needs. Accompanied by the popularization and deepening of computer technology, pace of the information construction has been accelerated, and increased of user aspirations and needs, MULW almost have all the resources and services of the library. MULW is not only the network service platform of the library, it is also the library image on the network. It is the extension and deepening of the development concept of library, including service concept, safety concept and culture concept. [1, 2] Therefore, the MULW needs not only to enhance its own practicability, and it needs to satisfy the user omni-directional demand to improve the user experience. This is the core connotation of MULWC. In order to better improve the MULWC, university librarians need to constant discovery and satisfaction user needs.

2. MULWC basic function demand

The basic function demand of MULWC, it means that users can browse the library website and get basic satisfaction. This is also the basic functions that the library website must have. In library website construction, make use of various functional modules of the library website to meet users' basic needs. It is the process of library website for enriching the content. To refine the various functional modules of the library website, and make each functional module has the basic information which user needs by browsing.

2.1. Exhibition function
This function can fully demonstrate the significance and function of the library, to make users fully understand the library resources and services. In the network environment, due to the rapid growth of information resources, users often cannot rely on their own power to quickly and accurately find the required information from the sea of information.[3] When users browse library website in the first time, the exhibition function can give a general idea of what basic services and resources library can bring. It can guide readers to confirm the correct direction in the sea of information, and search for the information they need timely, accurately and conveniently. Then users can get the basic information needs. Under the condition of understanding the various functions of the library website, when users logging in library website it is necessary to meet the requirements of library resources search, information service, book borrowing and loading, Information space appointment, etc.

2.2. Resources and Services using function
This function effectively communicates the library own resources and services to users. Clear target users and improve communication efficiency, so that users can get the resources and services of the library more quickly and accurately. Therefore, the library website needs to provide convenient resources and service utilization function for teaching and research, bring aided teaching to the students, and bring scientific research support to university staffs and researchers. In order to optimize this functionality, there is a higher requirement for the quality, access speed, retrieval speed, retrieval accuracy and update timeliness of library information resources.

2.3. Extended service function
This feature extends library services to the library website, it improves the service efficiency of library website and also saves users' time. The user is not required to go to the library, and also can enjoy the services provided by the library when using the library website. Keep the library offline service as much as possible on the library website, and become a combination of offline services and online services. At the same time, enhance the diversity of service terminals, and let this function module can be used more efficiently by users.

2.4. Resource rendering function
This function extracts and streamlines the library high-density database resources to the users, and let a large number of databases are presented in an orderly manner on some limited web pages. Good navigation can organize complex page contents, sort out the logic, and establish a concise and easy to understand column information structure.[4] In order to make it is convenient for users to use the library website database resources, the library website will make a bold distinction between Chinese and foreign language databases, and it is usually arranged in order according to the usage rate. In the lookup method of database, user can search by letter A to Z, search by the literature type provided by the database, search by subject area, search by database function, search the database name directly, etc. It is convenient for users to find the target database they want to use, and to better provide professional information resource service.

2.5. User education function
This function is designed to enhance the user's ability when they use the library and library website, to make them can skillfully use the library services and obtain the library resources, and to reflect the library intellectual value. User education, on the one hand, allows users to absorb knowledge about how to use the library, the essence is to take advantage of the stock knowledge in the society. And maybe generate new knowledge, thus, the value of stock knowledge can be increased.[5] The library website can easily provide library education for users, let the education mode combine online and offline to teach users how to use library resources and services.

2.6. Communication and interaction
This function enables the user's ideas to be conveyed to the library. It can not only effectively solve the problem of users, but also improve the service quality of the library. Library use website to communicate frequently with users, and understand what users need in communication. Library website builders need to

connect library and users closely on the library website, care about the users experience of using the site, and maintain a close contact with users.

3. MULWC security demand

The security demand of MULWC means that users can use the site in a relatively secure network environment. In the user security requirements experience, the library website need to make the users personal information will not be disclosed, to have a good confidentiality, to have a good background control, to make the library website server will not be hacked, to keep there is no illegal information or bad links on the library website, etc. These factors can affect the security level of the library website. The security requirement must have the library website security module construction. Library website builders need to ensure that the site can reach a high level of security usage and run in a secure environment which is the website security system (Figure 1).

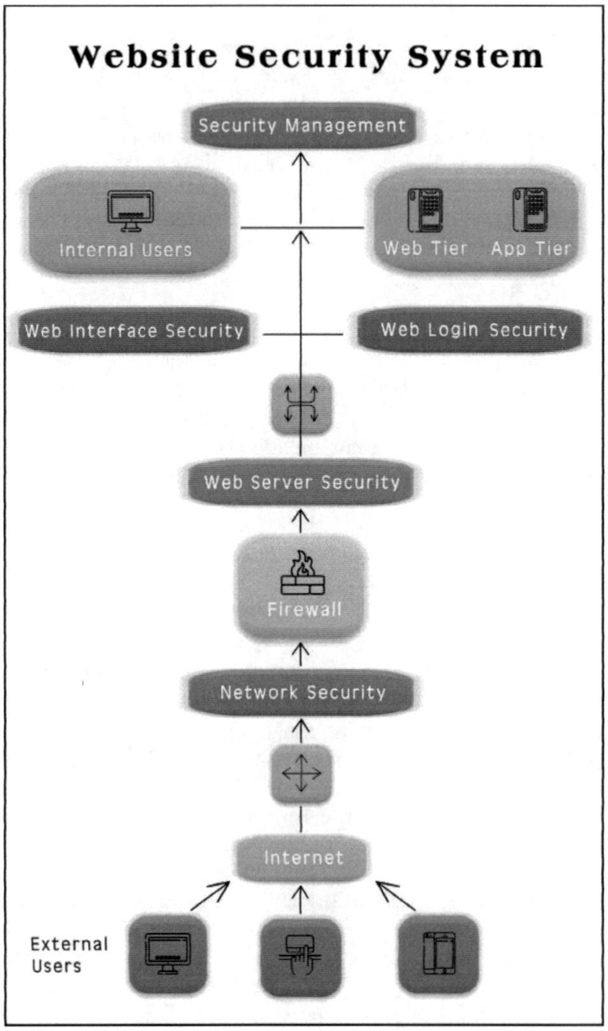

Figure 1. Website Security System

3.1. Network security

The CMULW is inseparable from network, and various network security vulnerabilities can affect the security of the library website and user experience. It is necessary to ensure network security, and also it is the basis of security requirement for users. Although network security threats exist objectively, but the risk can be controlled or avoided.[6] In the MULWC, rather than passively waiting for the emergence of network security issues, the library website builders should take the initiative to guard against it. On the one hand, the library website builders should actively protect the network of library websites, on the other hand, users also need to enhance their own network security awareness.

3.2. Library website server security

Library website server security management is one of the important contents of website security management. Website server management and maintenance can provide sufficient guarantee for the safe operation of the website.[7] The first is the protection of the server room, through a regulated system keep the library website safe. The second is to keep server communication is of high quality, ensure that qualified and stable transmission routes and network communication are maintained, make sure users get a comfortable browsing experience. Third, the web server equipment must have been secured to protect web applications from distributed denial of service attacks. Supports content filtering to prevent network attacks, such as SYN Flood, Teardrop, Ping-of-Death, Smurf and other malicious attacks. Furthermore, the server own security protection includes firewall protection, patches update, routine security configuration and maintenance, etc.

3.3. Library website interface security

The library website interface is convenient for maintainers to manage and maintain the website. Through the library website interface, website maintainers can monitor the status of the website at any time, or make maintenance adjustments to the website. In some case, the library website interface also has the possibility of being exploited by hackers. Website maintainers need to prevent irrelevant personnel from using the external interface of the website, such as hackers can use the remote desktop and other tools to access the library website server. The second is to pay attention to the access of the network protocol used by the library website and quick entry for developing website. Website management requires effective management and control of all interfaces to the library website, and need to ensure that the data in the network environment is complete and confidential.

3.4. Library website login authentication security

The login authentication system of library website is to protect user's privacy, personal information and browse information. On the other hand, it is to facilitate website management and monitor users browsing behavior. Library website login strictly unified authentication, and guarantee one person an account. When users browse the library, website encountered an error page, website maintainers can discover and solve problems in time. When users account exception occur, website maintainers can also detect and stop users inappropriate browsing behavior such as excessive downloading, unknown non-security links browsing, etc.

3.5. Library website security system

The safety of library website is inseparable from the effective safety management system. The first is to base on technical strength of library website to formulate the technical depth of system management. The second is to base on the risk requirements of the library website to formulate the risk prevention countermeasure that suit oneself actual situation. Library website builders need to make a combination of technical means and management measures to deal with security issues, and to form the security management system of the library website.

4. MULWC humanistic respect demand

The humanistic respect demand of the MULWC is the users who use the library website for a long time to produce a need for website dependencies. This dependence is pleasurable for users and is essential for daily work. Let users to make it a habit to use the library website, it is the pursuit that the MULWC must pay attention to.

4.1. Simple and beautiful design

Library websites should be accessible easily, and can embody the user-centered service ideas. Make sure library website fits most user's usage habits. The effective method is to adopt contracted and beautiful website design. It is the goal of website designers to achieve the best expression through simple and beautiful expression.[8] The main characteristic that contracted beautiful website design is to delete all unnecessary redundant design. Through professional and convenient navigation in this design, library website can quickly reflect the information required by users. Using library website can give users a relaxed feel.

4.2. Enhance user communication

MULWC should always respect the needs of users, and let users wishes be valued and satisfied. This kind of respect demand can be understood as the user can get timely solution when having opinions or suggestions, and maintain an interactive mode with the user for a long time. To this end, library website builders can set up a website interaction column and to establish an interactive communication platform with users online or offline. Users can interact with librarians at any time through the Internet and can truly become participants in the website construction, which will play an important role in promoting the improvement of service quality and website level.[9] This interaction can also use a variety of ways, such as conduct regular site satisfaction surveys, answer the hot questions with high attention of users on a regular basis, set up the director mailbox, etc.

4.3. Good website maintenance

The library website needs to be maintained to implement the user's ideas and suggestions. The MULWC should be improved and developed in the direction desired by the users. For the maintenance of computer websites, it is necessary not only to update the information displayed in the website in a timely manner, but also to control and manage the various items in the website.[10] So website maintenance is essential work. Good website maintenance can ensure the stable operation of the website, and can strengthen the website construction according to the reasonable suggestions of users. In the future work, through the maintenance of the website, the library can timely discover the new needs of users to improve and strengthen the MULWC.

5. Conclusion

The above analysis shows that the MULWC should meet the basic function demand, the security demand and the humanistic respect demand of users. The MULWC needs advanced science and technology, and also needs a kind of construction idea to meet the needs of users in all directions. The MULWC is not only satisfying the realization of user value, but also indirectly improving self-value. Users are the users of the library website, and are the participants and supervisors of the MULWC. The basic attitude and development path of MULWC needs to constantly discover the demands of users and try to meet these demands.

References

[1] Jinbo REN. The construction and maintenance of university library website[J]. *Journal of Library and Information Sciences in Agriculture*, 2005, (12), 133-142.
[2] Zhenjie PAN. On the user experience design of library website[J]. *Journal of Library and Information Sciences in Agriculture*, 2012, 24(06), 154-157.
[3] Zhiwei YU, Zhiyin Yang. Research on college library personalized information service under network environment[J]. *Agriculture & Technology*, 2008, 28(3), 185-189.
[4] Jianling QIU. An empirical study on improving the user experience of university library website[J]. *Journal of Library and Information Sciences in Agriculture*, 2017, 29(12), 105-109.
[5] Hong YU. The influence of reader education on library knowledge transfer and sharing[J]. *Library Journal*, 2014, 36(06), 1-3.
[6] ChunYuan WANG, Shanlin Yang, Yongjun LIU. Information network security event monitoring and response platform design[C]. *Hunan: proceedings of the National Academy of Computer Security Symposium* (Volume 22), 2007.
[7] Wenxue MA. Website maintenance and security management[J]. *China Management Informationization*, 2015, 18(04), 206.
[8] Zhenhua HU, Zongliang WANG, Zhirong WU. The pursuit of minimalism about the university library website style of thinking[J]. *Intelligence work Research of Shanghai University Library*, 2010(2), 1-4.
[9] Xuemei LI. Comparative analysis of the interactive columns of Chinese and foreign university libraries[J]. *Researches in Library Science*, 2016(4), 29-36.
[10] Chunjin XU. Maintenance technology Research of computer website[J]. *Electronic Technology and Software Engineering*, 2015(17), 23.

The Resolution of Gender Anaphora Reference with the Help of Kernel Trick Mechanisms

A A Pak, S Amirzhan, A A Ziyaden

Institute of Informational and Computational Technologies, Almaty, Kazakhstan

aa.pak83@gmail.com, amir.sanzhar@gmail.com, iamdenay@gmail.com

Abstract. An article represents the results of the experiments constructing of conventional anaphora resolution system for gender-balanced corpora GAP. The main idea behind the conducted set of experiments is taking advantage over the kernel trick mechanism of deep word vectorization for various semantic and syntax features. We will show that it has a positive impact on system performance versus the traditional approach.

1. Introduction

Coreference Resolution (CR) is a crucial step in the pipeline of many high-level applications of Natural Language Processing (NLP), namely fact extraction, machine translation, text summarization, etc. The task of CR begins its history from the first attempts to resolve the connection between pronoun and noun phrases [1], nowadays, such a task called Anaphora Resolution (AR). Subsequently, at Message Understanding Conference, some partial subproblems [2] were resolved. Traditionally, algorithms, as the first step, generated the set of mentions of entities. After that, each pair of pronouns and mention was classified with the help of a decision tree algorithm. At the end of processing the positively classified pairs were merged into a cluster, that represents itself the dedicated chain between connected contexts of mentions and noun phrases. In the paper [3] the modification of generative grammar was presented, namely «Government and Binding theory» (GB). The significant part of the theory was dedicated to syntax anaphora, i.e., analysis of links between pronoun and other parts of the speech.

The next epoch of coreference resolution begins with the conference "ConLL-2011/2012" and the commonly accessed task of modeling non-limiting coreferences; as a result, a new variety of approaches appeared [4]. Coreference is a link between mentions (noun phrases), which are related to the same entity. By analogy with GB, there is an approach that utilized multi filtration, where each level of filter setups or denies a link between words. At the paper [5] it was indicated that on the general noun phrase coreference task the learning approach achieves accuracy comparable to that of nonlinear approaches, moreover, instead of considering the task of AR as a local optimization problem, authors proposed new consideration of the task as global optimization among all pairs in the document.

There are a wide variety of manually marked resources and corpora. Nonetheless, most of them do not capture ambiguous pronouns in enough volume or diversity to accurately indicate the practical applicability of models. Furthermore, there is a lack of balanced corpora due to gender bias in existing corpora and systems favoring masculine entities. A new corpus was presented in order to overcome these shortcomings, a gender-balanced labeled corpus (GAP) of 8,908 ambiguous pronoun-name pairs sampled to provide diverse coverage of challenges posed by the real-world text. There is a range of baselines that demonstrates the complexity of the challenge; the best result is just a 66.9% F1 score [6].

IOP Conf. Series: Materials Science and Engineering **630** (2019) 012029 doi:10.1088/1757-899X/630/1/012029

2. Word embeddings: BERT

Word embeddings are a form of computational distributional semantics for determining a word's meaning "from the company it keeps" [7,8], i.e., the words it co-occurs. The most popular model is recently presented Bidirectional Encoder Representations from Transformers (BERT). Unlike other word embeddings [9,10] BERT is designed to pre-train deep bidirectional representations by jointly conditioning on both left and right context in all layers. As a result, the mentioned model suits well for a wide range of NLP tasks, such as question answering and coreference resolution and others [11].

BERT's model has several common features with ELMO and OpenAI GPT. It is a multi-layer bidirectional Transformer encoder based on the original realization described in [12]. Nowadays, the implementation of Transformers is so popular that we will omit the exhaustive model architecture description and refer readers to Vaswani et al. [12].

3. Kernel trick

A kernel trick is a mathematical tool that can be applied to any algorithm which consists of a dot product between two vectors. The main idea of the kernel trick is getting a better discriminability of samples' classes with the help of feature space extension. Kernel trick assumes that to compute the dot products of vectors in the higher dimensional space; a kernel function can be used directly with the lower-dimensional vectors. Therefore, every dot product can be replaced by a kernel function. A Kernel function has its area of definition in the original space and returns the dot product of the vectors in the enriched feature space, which can be produced with the help of simple algebraic operations over the original space with low dimension. The feature space is usually a higher dimensional space where data can be linearly separable. In order for a function to be a valid Kernel function, it should fulfill the "Mercer's theorem". According to the Mercers' theorem, every positive definite symmetric function can be seen as a kernel function.

4. Dataset codebook and extra features

The task is to identify the target of a pronoun within a text passage. The source text is taken from Wikipedia articles. In the dataset, there are labels of the pronoun and two candidate names to which the pronoun could refer. An algorithm should be capable of deciding whether the pronoun refers to name A, name B, or neither. The dataset is also available on the GAP Github Repo.

There are the following columns for analysis:

- ID - Unique identifier for an example (Matches to Id in output file format);
- Text - Text containing the ambiguous pronoun and two candidate names (about a paragraph in length);
- Text - Text containing the ambiguous pronoun and two candidate names (about a paragraph in length);
- Pronoun - The target pronoun (text);
- Pronoun-offset The character offset of Pronoun in Text;
- A - The first name candidate (text);
- A-offset - The character offset of name A in Text;
- B - The second name candidate;
- B-offset - The character offset of name B in Text;
- URL - The URL of the source Wikipedia page for the example;
- Extra features

$$\vec{w}_{ab} = \vec{a} \cdot \vec{b} \quad (1)$$

, where $\vec{a}$,$\vec{b}$ -vector representations of A and B candidates; $\cdot$ - element-wise multiplication.

$$\vec{w}_{pp} = \vec{p} \cdot \vec{p} \quad (2)$$

, where $\vec{p}$ -vector representations of pronoun; $\cdot$ - element-wise multiplications.

$$\vec{w} = \exp(\| \vec{w}_{pp} - \vec{w}_{ab} \|) \quad (3)$$

The following features increased the performance of the model: distances between A and B candidates versus pronoun; the flag of the determinant (his or her) usually links to a subject, syntax roles; kernel function as described above. There are 4000 samples with 6149-dimension size.

5. Models and Experiments

Our neural network is Feed-Forward Network(FFN) which consist 2 hidden layers of size 2048 with dropout 0.7 and 254 with dropout 0.2, activation function used 'relu'. Last layer is softmax - probabilistic distribution of classes, see figure 1.

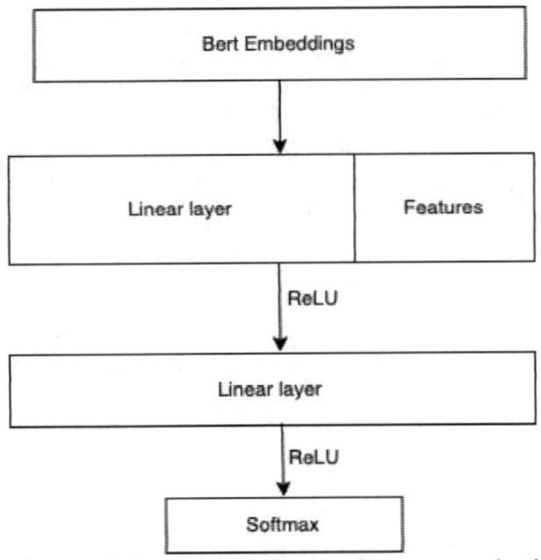

Figure 1. The architecture of anaphora resolution classifier.

6. Results

The task is to identify the target of a pronoun within a text passage. Together with various vectorization architectures, various additional features were tested. You can see the comparative analysis in Table 1, in the first column, the model name and additional features are given to improve the classification. The second column shows the error in the form of the cross-entropy value.

Table 1. Model features versus log-loss function

Model	Validation Loss
Base BERT	0.63
Large BERT	0.58
Large BERT + features	0.52
Large BERT + kernel trick + features	0.45
Large BERT + [18,20] layers + features	0.42
Large BERT + ALL	0.33

7. Discussion and Conclusion

Labeled corpora of texts for training and testing modern systems of natural language processing have had a statistical bias towards the masculine. In the present work, we considered the classical problem of resolving anaphoric connections as exemplified by the gender-balanced corpus of texts. The problem can be described as finding the chains of interrelated words in meaning from the stream of tokens, the syntax of noun groups and pronouns. It is worth noting the significant advantage of vectorization using BERT compared to traditional methods. In order to improve the model, we tested several additional features, namely the distance between noun groups and a candidate-pronoun in tokens; a sign of a linguistic determinant; other syntactic features; and the kernel function, which made the most significant contribution to the improvement of the model. A potential direction of development is testing new word vectorization models and expanding the corpus of gender-balanced texts.

Acknowledgements
We gratefully acknowledge financial support of Ministry of Education and Sciences, Republic of Kazakhstan (Grant AP05132760).

References

[1] Hobbs J R 1978 Resolving pronoun references Lingua, 44 (4), pp. 311-338
[2] Seventh Message Understanding Conference: Proceedings of a Conference Held in Fairfax,Virginia, USA, MUC 1998, April 29 - May 1, 1998. ACL 1998
[3] Chomsky N 1981 Lectures in Government and Binding. (Studies in generative grammar 9.) Dordrecht: Foris
[4] Pradhan S, Moschitti A, Xue N, Uryupina O, Zhang Y 2012 CoNLL-2012 shared task: Modeling multilingual unrestricted coreference in OntoNotes. In *Proceedings of the Sixteenth Conference on Computational Natural Language Learning* (CoNLL 2012), Jeju
[5] Soon W M, Ng H Y, Lim D C Y 2001 *A machine learing approach to coreference resolution of noun phrases. Computational Linguistic(special Issue on Computational Anaphora Resolution),27(4):521-544*
[6] Webster K, Recasens M, Axelrod V, Baldridge J 2018 Mind the GAP: A Balanced Corpus of Gendered Ambiguous Pronouns.
[7] Firth J R 1957 *Applications of General Linguistics. Transactions of the Philological Society*, 56: 1-14. doi:10.1111/j.1467-968X.1957.tb00568.x
[8] Amirzhan S, Pak A, Jaxylykova A B 2018 The estimation of stability of semantic space generated by word embedding algorithms. 1-5. 10.1109/iSAI-NLP.2018.8692814.
[9] Peters M, Neumann M, Iyyer M, Gardner M, Clark C, Lee K, Zettlemoyer L 2018 Deep contextualized word representations. In NAACL
[10] Radford A, Narasimhan V, Salimans T, Sutskever I 2018 Improving language understanding with unsupervised learning. Technical report, OpenAI
[11] Devlin J, Chang M-W, Lee K, Toutanova K 2018 Bert: Pre-training of deep bidirectional transformers for language understanding. arXiv preprint arXiv:1810.04805
[12] Vaswani A, Shazeer N, Parmar N, Uszkoreit J, Jones L, Gomez A N, Kaiser L, Polosukhin I 2017 Attention is all you need. CoRR, abs/1706.03762

IOP Conf. Series: Materials Science and Engineering **630** (2019) 012030 doi:10.1088/1757-899X/630/1/012030

Modelling of Thermal Effects on Aerospace Devices and Their Components

A V Golikov[1], M A Barulina[1], E V Pankratova[1] and O V Golikova[1]

[1]Laboratory of the Analysis and Synthesis of Dynamic Systems in Precision Mechanics, Institute of Precision Mechanics and Control, Russian Academy of Sciences, 24 ul. Rabochaya, Saratov, Russia, 410028

marina@barulina.ru

Abstract. Thermal effects on aerospace devices (sensors and systems) directly affect their performance. Thermal influences are one of the main sources of sensor's errors. The study of the effect of the thermal influences on the devices should be carried out for each device at the design stage. This work is especially crucial for high-precision devices or devices intended to be used in hostile environments like space. The paper proposes a classification and mathematical expressions of temperature influences that affect different kinds of aerospace devices and their electronic components during their operation. The algorithms developed by the authors allow modelling, for example, the change in the ambient temperature of components of a spacecraft navigation system when it is moving in circular or elliptical orbits, to simulate a cyclogram or stochastic of a temperature changing. The usage of the suggested types of thermal influences and relevant mathematical models can help to reduce the development costs of software for modelling and studying thermal processes in different classes of aerospace devices and speed up the process of its development.

1. Introduction

Thermal effects are one of the factors that have a significant impact on the accuracy and stability of sensors and systems [1-14]. Such influences can be critical for inertial devices (sensors, units, systems) used in hostile conditions, for example, in outer space. Therefore, the study of the influence of various thermal effects on such devices should be carried out at the stage of their design. Moreover, all the disturbing temperature factors which can arise in real operating conditions of the device should be taken into account for adequate numerical simulation of thermal processes in it.

This work aims to classify and to formulate the mathematical expressions of the temperature influences which can be during the operation of various kind of aerospace devices, sensors, and electronic components.

2. The main types of thermal effects

The disturbing thermal effects which can be during the operation of the devices in various environmental conditions can be divided into six types:

2.1. Constant

The thermal influence can be described by the following law (Figure1a):

$$T_c(t) = \text{const.} \tag{1}$$

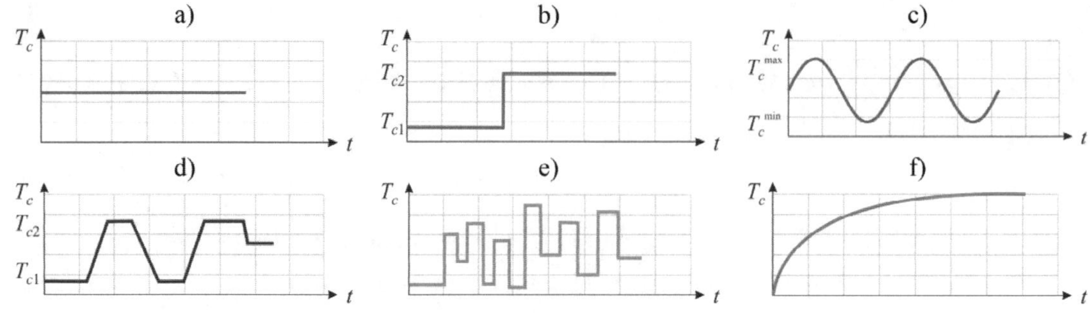

Figure 1. Types of thermal influence:
a – constant; b – stepped; c –harmonic; d – cyclogram; e – stochastic; f – free

2.2. Stepped

The temperature abruptly changes from T_{c1} to T_{c2} at the given time t_1 (Figure 1b):

$$T_c(t) = \begin{cases} T_{c1}, t \in [t_0, t_1] \\ T_{c2}, t \in [t_1, t_n] \end{cases} \quad (2)$$

2.3. Harmonic

The mathematical *expression of this kind of thermal influence (Figure 1c):*

$$T_c(t) = \frac{1}{2}\left(T_c^{\max} - T_c^{\min}\right)\sin(\omega t) + \frac{1}{2}\left(T_c^{\max} + T_c^{\min}\right), \quad (3)$$

where $T_c^{\max}$, $T_c^{\min}$ - maximum and minimum values of temperature accordingly.

Such law of temperature changes can be used for imitation, for example, a change in the ambient temperature of the elements of the navigation system of a spacecraft in a circular orbit.

2.4. Cyclogram

Thermal influence is changed according to defined thermal cyclogram (Figure 1d):

$$T_c(t) = (T_{c,i+1} - T_{c,i})\frac{t - t_{i-1}}{t_i - t_{i-1}} + T_{c,i}\text{ , } i = 1..n, \quad (4)$$

where i – cycle number; $T_{c,i}$ - temperature value in i-th cycle; t_i - i-th cycle start time.

The proposed law $T_c(t)$ allows simulating any cyclogram of a temperature change. This law of temperature influence is used, for example, for temperature specifying in tests of the devices and their components in a heat chamber. An example of using this law is shown in Figure 2. There are the numerical simulation results, and experimental data of temperature changing in a fiber-optic gyroscope as part of a strapdown inertial navigation system placed in a heat chamber is shown in Figure2.

2.5. Stochastic

Thermal influence is changed randomly. In this case, the mathematical law of temperature can be defined as follows (Figure 1e).

$$T_c = M + \sigma\left(\left(\left((a_9 r^2 + a_7)r^2 + a_5\right)r^2 + a_3\right)r^2 + a_1\right)r, \quad r = \frac{\sum_{i=1}^{12}\delta_i - 6}{4}, \quad (5)$$

where M - the mean; σ - standard deviation; $a_1, a_3, a_3, a_5, a_7, a_9$ – weight coefficient, $\delta_i \in [0,1]$ - random function.

2.6. Free

If thermal influence cannot be defined by any of expressions (1)-(6) or their combination, it should be described manually by investigators. It can be any function or series (Figure 1f):

$$T_c = f(t), \tag{6}$$

where $f(t)$ - linear, exponential or another kind of functions of time dependence $f(t)$:

$$f(t) = at + b \text{ или } f(t) = A(1 - e^{-at}), \tag{7}$$

where a, b - some constant parameters.

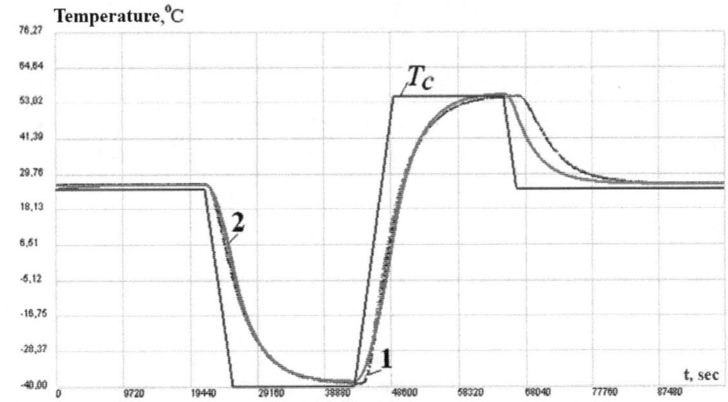

Figure 2. Current temperatures in SINS gyroscopes when modelling thermal processes: 1 – experiments data; 2 – simulation result; *Tc* – thermal influence

Expressions (1) - (6) are algorithms for modelling of various types of thermal influence. These expressions can be used in the development of software for modelling the influence of thermal effects on the object under study.

3. Algorithms for thermal influence simulation

Formulas (1) - (7) can be used as continuous functions or discrete sets of values in multi-purpose simulation software packages (for example, ANSYS, Code-Aster, MATLAB, etc.), as well as in specialized developed software and libraries for specifying the functions of thermal effects on the object or its structural elements.

The specialised software for thermal effects modelling can be developed with taken into account formulas (1)-(7) and based on the modified heat balance method (TEB).

TEB is a numerical method which looks very similar to the finite element method. So, the model is subdividing into smaller parts – thermal elements. The elements can contain sources and(or) sinks of heat. They can connect to other elements and the environment. The environment can have a temperature that varies according to some law.

The primary system of equations in TEB has the form [1,2]:

$$T_i(t + \Delta t) = \left[1 - \frac{\Delta t}{c_i}\left(\sum_{j=1}^{N} q_{ij} + q_{ic}\right)\right] T_i + \frac{\Delta t}{c_i}\left(\sum_{j=1}^{N} q_{ij} T_j + q_{ic} T_{ci} + Q_i\right), \tag{8}$$

where $T_i(t)$, $T_i(t + \Delta t)$ (i = 1,…,M) – the temperature of the i-th thermal element at time moment t and next moment $t + \Delta t$; c_i - heat capacity of the i-th thermal element; q_{ij} – characteristic of thermal connection between i-th and j-th thermal element (j=1,…,N); q_{ic} – characteristic of thermal connection between i-th thermal element and the environment; T_{ci} – ambient temperature; Q_i – power of source of heat or drain; M – total amount of TE in the model; N – amount of thermal element, which has thermal contact with i-th thermal element; Δt – calculation step.

The parameter T_{ci} in (8) is a function of time and should be defined by one of the expressions (1)-(7) or their combinations.

4. Software and numerical simulation

An example of the developed software for studying the effect of thermal influence on the angular velocity measuring unit with fibre-optic gyroscopes is shown in Figure 3

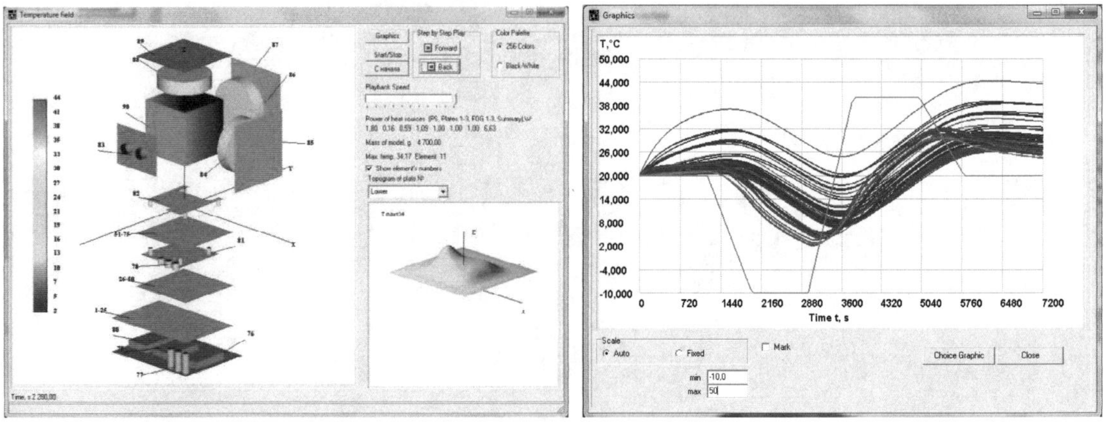

Figure 3 Specialized software for studying the effect of thermal influence

Numerical experiments have shown the sufficiency of the formulated types of thermal effects for modelling various temperature perturbations on the object under study.

5. Conclusion

The main types of thermal influences and the mathematical expression for them were formulated. These mathematical expressions can be used for numerical simulation and study of non-uniform non-stationary three-dimensional temperature fields with the help of well-known multipurpose or special software.

The suggested algorithms are adapted and ready for implementation in specialised or multipurpose software. So these algorithms can imitate various parameters of thermal effects, as well as other parameters of the ambient environment. For example, a change in the ambient temperature of the elements of the navigation system placed into a spacecraft can be simulated by the harmonic law when a spacecraft is moving in a circular or elliptical orbit, and by cyclogram in case of more complicated motion of the spacecraft. The parameters, such as weightlessness and vacuum, can also be taken into account.

The usage of the suggested types of thermal influences and relevant mathematical models can help to reduce development costs and speed up the process of development of software for modelling and studying thermal processes in different classes of aerospace devices.

Acknowledgements

The research was carried out within the state assignment of the Ministry of Science and Higher Education of the Russian Federation (theme No. AAAA-A18-118042790040-0)

References

[1] Dzhashitov V E, Pankratov V M 2005 Aerospace and marine instrument-making sensors, devices and systems under thermal actions *CSRI "Elektropribor"* 404 p

[2] Barulina M A, Golikov A V, Pankratov V M, Efremov M V 2018 Numerical study of three-dimensional unsteady thermal fields of aerospace high-precision sensors and system *Nauchnoe priborostroenie* 28(3), pp.14-23

[3] Dzhashitov V E, Pankratov V M, Barulina M A 2013 Mathematical models of thermal stress-strain state and scale factor error of fiber optic gyro sensors *Journal of Machinery Manufacture and Reliability* Vol. 42 No. 2. pp. 124–131.

[4] Teng F, Xu X, Zhang Z, An M, Zhang C 2015 Modeling and reduction of thermally induced non-reciprocal error indifferential interference fiber optic gyroscope sensing coil *Optic* Vol. 126 No. 13 pp. 1295–1299.

[5] Golikov A V, Pankratov V M, Efremov M V 2018 Analysis of Temperature Fields in Angular Velocity Measurement Units on Fiber-Optic Gyros *Gyroscopy and Navigation* Vol. 9 No. 2, pp. 116–123

[6] Li M, Zhao X 2010 A Novel 3-D Model for Thermal Transient Effects in Fiber Gyro Coils *Symposium Gyro Technology*, 21-22 September, Karlsrue, Germany.

[7] Quatraro E, Pizzarulli A, Catasta M, Crescenti G, Spinozzi E, Cingolani A 2011 High Performance FOG for Non Temperature Stabilized Environment *Inertial Sensors and Systems – Symposium Gyro Technology*, 20-21 September, Karlsrue, Germany.

[8] Schadt F, Mohr F. 2011 Error Signal Formation in FOGs Through Thermal and Elastooptical Environment Influence on the Sensing Coil *Inertial Sensors and Systems – Symposium Gyro Technology*, 20-21 September, Karlsrue, Germany.

[9] Shupe D M 1980 Thermally induced non-reciprocity in the fiber-optic interferometer *Appl. Opt.* Vol. 19 No. 5 pp. 654-655

[10] Dzhashitov V E, Pankratov V M, Golikov A V, Nikolaev S G, Kolevatov A P, Plotnikov A D, Koffer K V 2014 Hierarchical thermal models of FOG-based strapdown inertial navigation system *Gyroscopy and Navigation* Vol. 5, Issue 3, pp 162–173

[11] Zhang Y, Gao Z, Wang G, Gao W 2014 Modeling of thermal-induced rate error for FOG with temperature ranging from −40 °C to 60 °C *Photonics Technology Letters* Vol. 26, No 1. pp. 18–21

[12] Yonggang Z et all. 2014 Modeling of thermal-induced rate error for FOG with temperature ranging from –40 to +60 °C *IEEE Photonics Technology Letters*, vol. 26, no. 1, pp. 305–310

[13] Lefevre H 1993 The Fiber-Optic Gyroscope *Norwood, MA: Artech House* p. 24.

[14] Dulnev G N, Parfenov V G, Sigalov A V 1990, Methods for calculating the thermal regime of devices. *Moscow: RADIO ISVYAZ'*, 312 p. (In Russian)

IOP Conf. Series: Materials Science and Engineering **630** (2019) 012031 doi:10.1088/1757-899X/630/1/012031

Analysis of emerging disturbing factors on the descent vehicle with inflatable mechanical devices during the descent stage

Vsevolod V. Koryanov [1], **Sun Huanyu** [1, 2]

[1] Bauman Moscow State Technical University, 5, b1, 2-nd Baumanskaya Street, Moscow, Russian Federation, 105005
[2] Harbin Institute of Technology, 92 West Dazhi Street, Nan Gang District, Harbin, People's Republic of China, 150001

vkoryanov@mail.ru

Abstract. The movement of asymmetric descent vehicles, in contrast to the movement of symmetrical landing vehicles, has a number of features. The main of them consist in the emergence of such dynamic phenomena as progressive self-rotation, vibrational-rotational resonances, nutation-precession instability, auto-oscillations, and others. In this paper, the entire landing trajectory is divided into main stages of movement, and then an analysis of some of the most unfavourable variants of movement at each stage is given. The conclusion is about the worst options. This must be taken into account when designing such descent vehicles with inflatable structures.

1. Introduction

At present, a large number of space missions are being designed. It is increasingly important to perform tasks to ensure that goods are returned to the small automated station from track or space. This may be the return of a sample in an interplanetary mission, or it may be the return of cargo from Earth orbit. For such missions, it is necessary to make a reliable landing for the landing vehicle. One of the most important stages in the movement of a spacecraft is the final stage - the landing of a spacecraft on the surface of the planet. The difficulty lies in the fact that it is necessary to reduce the speed from huge values during entry into the atmosphere to permissible values when landing on the surface of the planet. Another important limitation is the weight limit and the geometric parameters of the spacecraft.

Thus, the option to decelerate the space descent vehicle is to use inflatable braking devices. One of such projects in which inflatable braking devices are used is the MetNet project and its continuation RITD project [1-5]. (Figure 1). This is an aerospace recycling system that provides lift or resistance using aerodynamic shapes formed by aeration during reentry [6, 7].

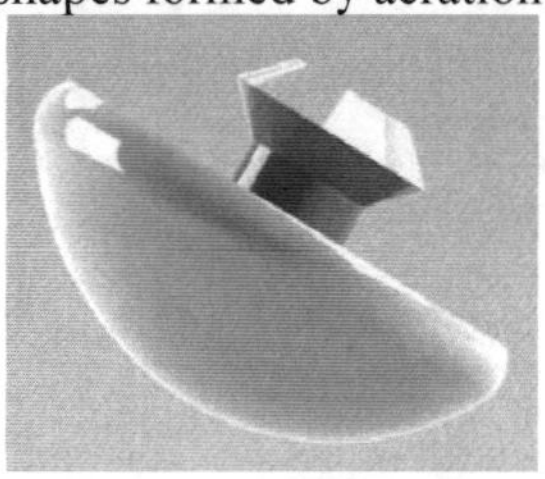
I

II

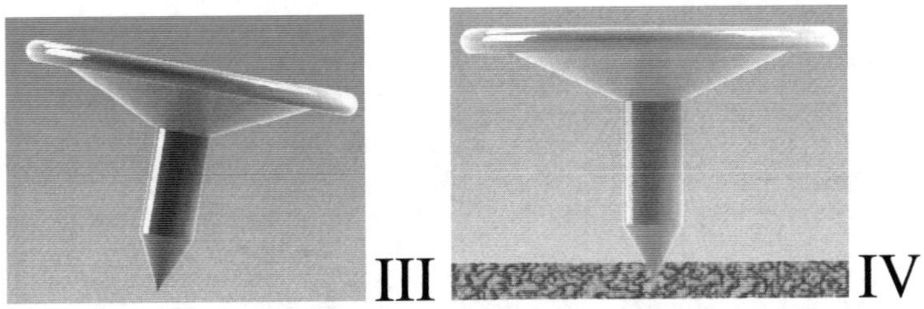

Figure 1. Descent scheme of landing vehicle movement

Figure 1 shows the main stages of the movement: I - Atmosphere entry. Opening primary inflatable braking device; II - Entering the additional inflatable braking device; III - Braking with the additional inflatable braking device; IV - Landing vehicle touch the ground.

Compared with the traditional rigid pneumatic deceleration method, the pneumatic reducer has two functions in the reentry process: the heat protection function of the reentry section, the aerodynamic thermal load subjected to the hypersonic flow when entering the atmosphere, and the pneumatic deceleration function. The speed requirement is achieved by aerodynamic deceleration. Because of its light weight, flexibility, high strength and high temperature resistance, it can be folded into a small volume for easy loading. At the same time, it has a large resistance windward area after deployment, which can provide effective deceleration effect [8].

This paper is devoted to the four moments of the landing process. When the center of mass of the pneumatic deceleration device is displaced, the load of the inflatable device will change greatly, and it is necessary to analyze the influence of this change and the resulting imbalance of force on the inflatable device.

2. Theoretical part for the descent into the atmosphere

In this section, we will study the forces of landing vehicles in the atmosphere, allowing for small asymmetry [2]. In order to brake into the atmosphere, we will consider landing vehicles with pneumatic brakes, and two of them: the main brake for the upper atmosphere and the additional brake for the lower atmosphere. During the descent, the landing vehicle and its braking device are affected by important disturbances due to the external environment. Therefore, due to its flexibility, the inflatable device may be deformed and the movement may become unstable [9, 10]. Materials of other scientist who work in the field of flexible structures are presented here [11]. In addition, there are a number of articles on deployable structures in space that are also of interest to our research [12, 13].

We will use a method to analyze how the design parameters and aerodynamic coefficients affect the asymmetry from the longitudinal axis of the vehicle's velocity vector.

We can write:

$$\dot{\omega}_x = \frac{1}{2}\Big[J_{xy}\cdot(\dot{\omega}_y - \omega_x\cdot\omega_z) + J_{xz}\cdot(\dot{\omega}_z + \omega_x\cdot\omega_y) + J_{yz}\cdot(\omega_y^2 - \omega_z^2) +$$

$$+(J_y - J_z)\cdot\omega_y\cdot\omega_z + q\cdot S\cdot l\cdot\left(m_{x_0} + m_x^{\bar{\omega}_x}\cdot\frac{\omega_x\cdot l}{V} + C_y\cdot\frac{\Delta z}{l} - C_z\cdot\frac{\Delta y}{l}\right)\Big],$$

Full equations are given in [1, 2].

In the future, we will change the values of the shift of the center of mass in order to obtain the most optimal variant from the point of view of the arising force factors.

Thus, taking into account the asymmetry of the descent vehicle with a device using an inflatable material occurs through a shift in the center of mass ΔCM in Figure 2.

The main idea is that there is a shift in the center of mass of the inflatable structure (Figure 2).

IOP Conf. Series: Materials Science and Engineering **630** (2019) 012031 doi:10.1088/1757-899X/630/1/012031

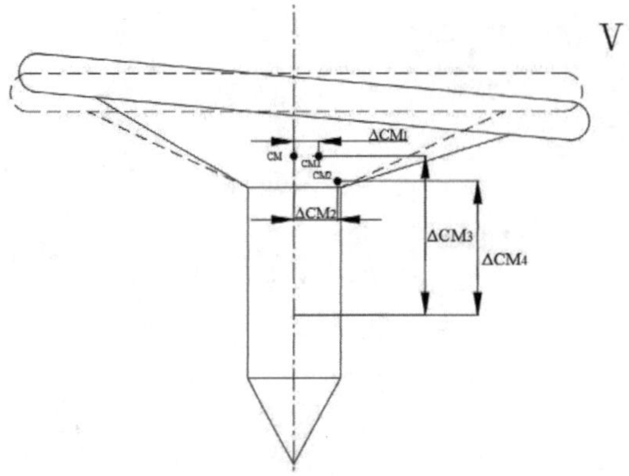

Figure 2. Centroid displacement of the inflatable device

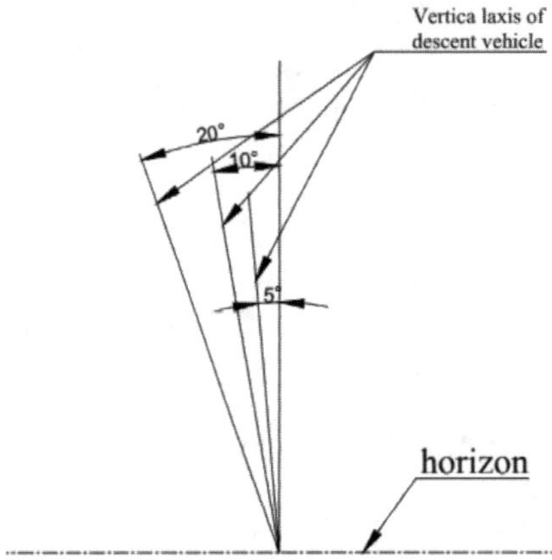

Figure 3. The slope of the longitudinal axis of the descent vehicle and the ground

3. Modelling part and analysis

We use software "Ansys" to solve this problem. We will focus on three moments in the landing process: heights of 60 kilometers, 20 kilometers and close to the ground.

We considered several options for the landing process: First, we believe that the vertical axis of the landing vehicle can be 0 degrees, 10 degrees or 20 degrees (Figure 3). Then, for each moment, we simulated the inflatable device to be offset by 5 degrees on one side. Finally, we consider the following data:

Speed to the ground (vertical speed component): Height 60 km: 2000 m/s, Height 20 km: 700 m/s, Height 0 km: 35 m/s.

We use the method of control points (Figure 4). Select seven points and observe the force at each point.

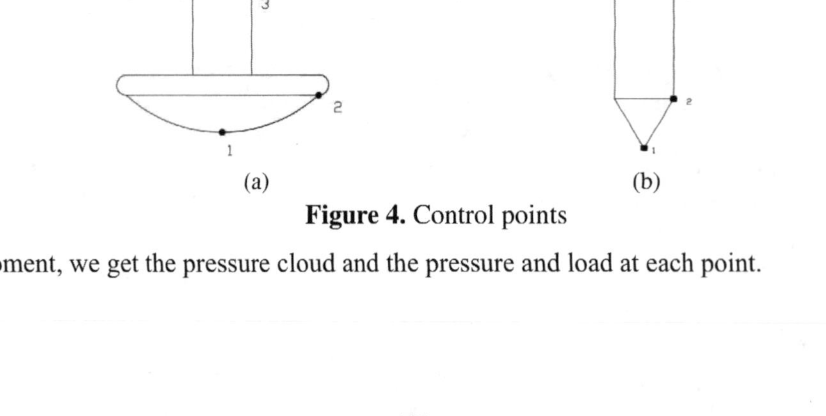

(a) (b)

Figure 4. Control points

For each moment, we get the pressure cloud and the pressure and load at each point.

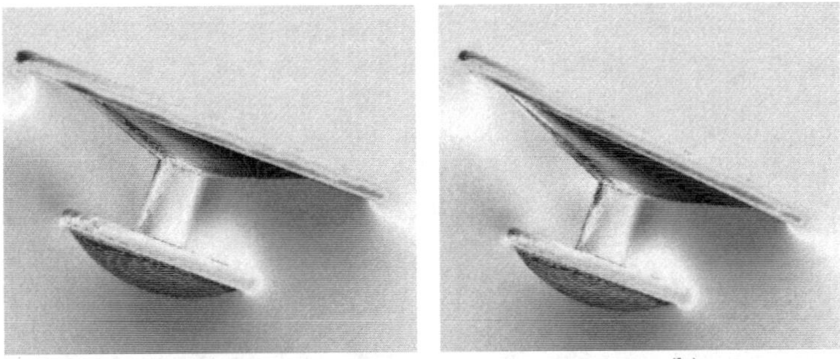

(a) (b)

Figure 5. The second stage of pressure cloud map when the angle of attack is 20 degrees (a) and with asymmetry 5 degrees (b)

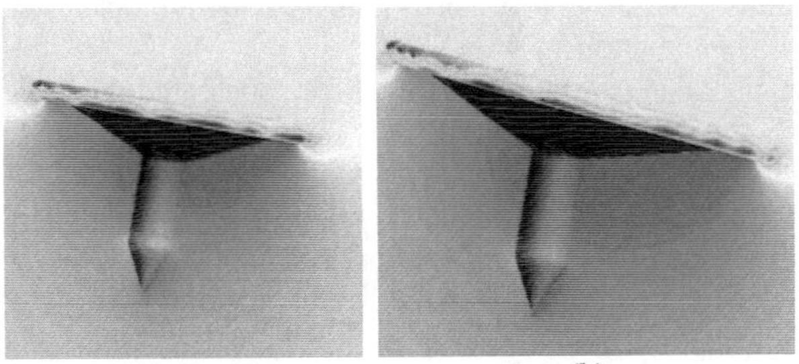

(a) (b)

Figure 6. The third stage of pressure cloud map when the angle of attack is 10 degrees (a) and with asymmetry 5 degrees (b)

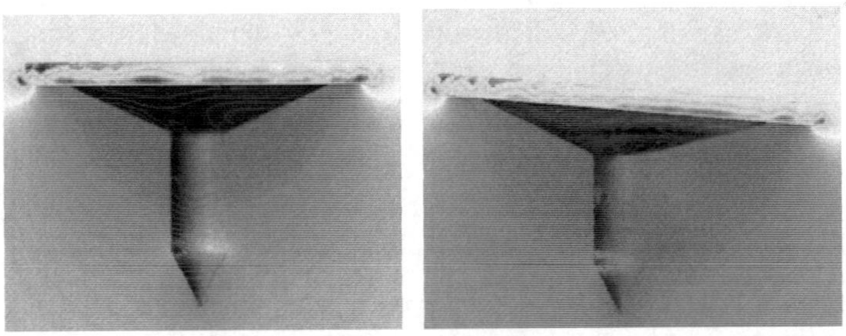

(a) (b)

Figure 7. The fourth stage of pressure cloud map when the angle of attack is 0 degrees (a) and with asymmetry 5 degrees (b)

Let us analyze the data obtained, which are graphically shown in Figures 5-7.
At the second stage of the movement, the calculation results of which are shown in Figure 5, we can conclude that the maximum pressure at point 1 in the absence of asymmetries and in the presence of asymmetries.
At the third stage of the movement, the calculation results of which are shown in Figure 6, we can conclude that the maximum pressure at point 4 in the absence of asymmetries and at point 3, taking into account the presence of asymmetries.
At the fourth stage of the movement, the calculation results of which are shown in Figure 7, we can conclude that the maximum pressure at point 3 in the absence of asymmetries and at point 4, taking into account the presence of asymmetries.
The most loaded stage of movement is stage number 3.

Thus, we can conclude that at various stages of movement the maximum values are reached at various control points of the structure, thus, we can conclude that the entire structure of the descent vehicle carries heavy loads.
The pressure and load on the inflatable structure increases due to the deformation of the inflatable structure.
From these values, we find that asymmetry greatly increases the loads acting on the descent vehicle, and disrupts the stress balance.

4. Conclusion

As a result of the research performed in this article, the simulation of the movement of the descent vehicle with an inflatable device has been performed. The shape asymmetry was modeled through the center of mass shift. This approach allowed us to simulate the arising asymmetries in the shape of the descent vehicle with an inflatable braking device. Different values of the center of mass deviation were considered. Received the most dangerous variants of deviation. The obtained pictures of the occurrence of forces and pressure for some of the most dangerous options of movement with an asymmetric flexible inflatable braking device.
Conclusions about the most dangerous options must be considered when designing such spacecraft in the future.

References

[1] V.V. Koryanov. Method of calculating the descent of the spacecraft in the atmosphere using technology adaptation landing in different environmental conditions, IAC-14-C2.3.4, 65th Astronautical Congress, Toronto, Canada, 2014, 29 September – 03 October.

[2] V.V. Koryanov, V.P. Kazakovtsev. Dynamics of angular motion of landing vehicle in Martian atmosphere with allowance for small asymmetries. *International Journal of Mechanical Engineering and Robotics Research*, Volume 7, Issue 4, 1 July 2018, Pages 385-391.

[3] Jyri Heilimo, Ari-Matti Harri, Sergey Aleksashkin, Vsevolod Koryanov, Ignacio Arruego, Walter Schmidt, Harri Haukka, Valery Finchenko, Maxim Martynov, Boris Ostresko, Andrey Ponomarenko, Viktor Kazakovtsev, Susanna Martin, and Tero Siili. RITD - Adapting Mars Entry, Descent and Landing System for Earth. *Geophysical Research Abstracts* Vol. 16, EGU2014-5506-1, 2014, EGU General Assembly 2014

[4] The Mars MetNet Mission. Concept Drawing. http://fmispace.fmi.fi/old-metnet/index.php?id-100, (accessed: 05.02.2019).

[5] Pichkhadze K.M., Finchenko V.S., Aleksashkin S.N., Ostreshko B.A. Transformable vehicles descending in the atmospheres of the planets, *Herald Federal State Unitary Enterprise Scientific and Production Association named after Lavochkin*, 2. 2015, 4-13.

[6] Michele Iacovazzo, Valerio Carandente, Raffaele Savino, Gennaro Zuppardi. Longitudinal stability analysis of a suborbital re-entry demonstrator for a deployable capsule. *Acta Astronautica. Volume 106*, January–February 2015, pp. 101–110.

[7] Xia Gang ,Qin Zizeng ,Zhang Xiajin. Development status of inflatable thermal shield technology. *Missiles and Space Vehicles* , 2002 ; (1) :19～24

[8] Li Shuang, Jiang Xiuqiang. Review and Prospect of Decelerator Technologies for Mars Entry. *Hangkong Xuebao/Acta Aeronautica et Astronautica Sinica* 36(2), 2015, 36(2): 422r440.

[9] Canuto E. et al. Planetary landing: Modelling and control of the propulsive descent //*Proceedings of the 31st Chinese Control Conference*. – IEEE, 2012. – C. 7309-7316.

[10] Cho D. H., Kim D., Leeghim H. Optimal lunar landing trajectory design for hybrid engine //*Mathematical Problems in Engineering*. – 2015. – T. 2015.

[11] V.V. Leonov. The design features of inflatable large-scale mirror concentrators for space high-temperature solar power plant. *Proceedings of the international astronautical congress*, IAC 67, Making Space Accessible and Affordable to All Countries. 2016.

[12] V.S. Zarubin, V.N. Zimin, G.N. Kuvyrkin. Temperature Distribution in the Spherical Shell of a

Gauge-Alignment Spacecraft. *Journal of Applied Mechanics and Technical Physics* (2017) 58. pp. 1083-1090. https://doi.org/10.1134/S0021894417060141.

[13] Nerovny, N.A., Zimin, V.N. Light radiation pressure upon a wrinkled membrane - Parametrization of an optically orthotropic model. *Journal of Physics: Conference Series Volume* 991, Issue 1, 13 April 2018, Article number 01206 25th International Conference on Topical Problems of Continuum Mechanics, TPCM 2017; Tsakhkadzor; Armenia; 2 October 2017.